Lecture Notes in Mechanical Engineering

For further volumes:
http://www.springer.com/series/11236

About this Series

Lecture Notes in Mechanical Engineering (LNME) publishes the latest developments in Mechanical Engineering - quickly, informally and with high quality. Original research reported in proceedings and post-proceedings represents the core of LNME. Also considered for publication are monographs, contributed volumes and lecture notes of exceptionally high quality and interest. Volumes published in LNME embrace all aspects, subfields and new challenges of mechanical engineering. Topics in the series include:

- Engineering Design
- Machinery and Machine Elements
- Mechanical Structures and Stress Analysis
- Automotive Engineering
- Engine Technology
- Aerospace Technology and Astronautics
- Nanotechnology and Microengineering
- Control, Robotics, Mechatronics
- MEMS
- Theoretical and Applied Mechanics
- Dynamical Systems, Control
- Fluid Mechanics
- Engineering Thermodynamics, Heat and Mass Transfer
- Manufacturing
- Precision Engineering, Instrumentation, Measurement
- Materials Engineering
- Tribology and Surface Technology

Jürgen Gausemeier · Franz Josef Rammig
Wilhelm Schäfer · Walter Sextro
Editors

Dependability of Self-optimizing Mechatronic Systems

Springer

Editors
Jürgen Gausemeier
Product Engineering
Heinz Nixdorf Institute
University of Paderborn
Paderborn
Germany

Wilhelm Schäfer
Software Engineering
Heinz Nixdorf Institute
University of Paderborn
Paderborn
Germany

Franz Josef Rammig
Design of Distributed Embedded Systems
Heinz Nixdorf Institute
University of Paderborn
Paderborn
Germany

Walter Sextro
Chair of Mechatronics and Dynamics
University of Paderborn
Paderborn
Germany

ISSN 2195-4356 ISSN 2195-4364 (electronic)
ISBN 978-3-642-53741-7 ISBN 978-3-642-53742-4 (eBook)
DOI 10.1007/978-3-642-53742-4
Springer Heidelberg New York Dordrecht London

Library of Congress Control Number: 2013956528

Printed on acid-free paper

Springer is part of Springer Science+Business Media (www.springer.com)

Preface

Intelligent technical systems offer the tempting possibility to create products that are far more capable than today's mechatronic systems. However, the system-inherent intelligence comes with increased system complexity. This leads to an increased risk for system-inherent faults, which might have devastating effects such as a complete breakdown or might even compromise the safety of users. Yet at the same time the new capabilities can be used to actively compensate for faults. Among these intelligent technical systems are self-optimizing systems, which are able to adapt their behavior based on user demands, the current situation or the system state itself.

Within the Collaborative Research Center (CRC) 614 "Self-Optimizing Concepts and Structures in Mechanical Engineering" at the University of Paderborn, we devoted our research to the development of these self-optimizing systems. They are based on mechatronics, which is formed by close collaboration of mechanical, electrical and software engineering. To make a system intelligent, the software engineering aspects gain further significance and are enhanced by advanced control engineering and mathematical methods. One result of our work is the accompanying book "Design Methodology for Intelligent Technical Systems", which introduces our proposed development process for self-optimizing systems. This process needs be accompanied by dedicated methods to assure or increase system dependability as early as possible in order to avoid costly design changes. We found that this augmenting process is so important yet also challenging that we decided to separate all related works into this book. It was written by the members of the CRC 614's working group "Safety and Stability" and coordinated by Tobias Meyer, to whom I'd like to express my sincerest gratitude.

The results of the CRC 614 and the book at hand present a major milestone towards the development of dependable intelligent systems. We hope that it inspires you to create tomorrow's products!

Your CRC-Team

Paderborn,
October 2013

Prof. Dr.-Ing. Jürgen Gausemeier
Speaker of the Collaborative Research Center 614

Acknowledgements

This book was written within the context of the Collaborative Research Center 614 "Self-Optimizing Concepts and Structures in Mechanical Engineering", which has been funded by the Deutsche Forschungsgemeinschaft (DFG).

Contents

List of Contributors

Prof. Dr.-Ing. habil. Wilhelm Dangelmaier
Heinz Nixdorf Institute, University of Paderborn, Business Computing, especially CIM, Fuerstenallee 11, D-33102 Paderborn, Germany,
e-mail: wilhelm.dangelmaier@hni.uni-paderborn.de

Prof. Dr. Michael Dellnitz
Chair of Applied Mathematics, University of Paderborn, Warburger Straße 100, 33098 Paderborn, Germany,
e-mail: dellnitz@uni-paderborn.de

Rafal Dorociak
Heinz Nixdorf Institute, University of Paderborn, Product Engineering, Fuerstenallee 11, D-33102 Paderborn, Germany,
e-mail: rafal.dorociak@hni.uni-paderborn.de

Kathrin Flaßkamp
Computational Dynamics and Optimal Control, University of Paderborn, Warburger Straße 100, 33098 Paderborn, Germany,
e-mail: kathrinf@math.uni-paderborn.de

Prof. Dr.-Ing. Juergen Gausemeier
Heinz Nixdorf Institute, University of Paderborn, Product Engineering, Fuerstenallee 11, D-33102 Paderborn, Germany,
e-mail: juergen.gausemeier@hni.uni-paderborn.de

Stefan Groesbrink
Heinz Nixdorf Institute, University of Paderborn, Design of Distributed Embedded Systems, Fuerstenallee 11, 33102 Paderborn, Germany,
e-mail: stefan.groesbrink@hni.uni-paderborn.de

Philip Hartmann
Heinz Nixdorf Institute, University of Paderborn, Business Computing, especially CIM, Fuerstenallee 11, D-33102 Paderborn, Germany,
e-mail: philip.hartmann@hni.uni-paderborn.de

Christian Heinzemann
Heinz Nixdorf Institute, University of Paderborn, Software Engineering Group, Zukunftsmeile 1, 33102 Paderborn, Germany,
e-mail: c.heinzemann@uni-paderborn.de

Christian Hölscher
Design and Drive Technology, University of Paderborn, Pohlweg 47-49, 33098 Paderborn, Germany,
e-mail: c.hoelscher@uni-paderborn.de

Peter Iwanek
Heinz Nixdorf Institute, University of Paderborn, Product Engineering, Fuerstenallee 11, D-33102 Paderborn, Germany,
e-mail: peter.iwanek@hni.uni-paderborn.de

Jan Henning Keßler
Heinz Nixdorf Institute, University of Paderborn, Control Engineering and Mechatronics, Fuerstenallee 11, D-33102 Paderborn, Germany,
e-mail: jan.henning.kessler@hni.uni-paderborn.de

Dr. Bernd Kleinjohann
Heinz Nixdorf Institute, University of Paderborn, Design of Distributed Embedded Systems, Fuerstenallee 11, 33102 Paderborn, Germany,
e-mail: bernd@c-lab.de

Dr. Lisa Kleinjohann
Heinz Nixdorf Institute, University of Paderborn, Design of Distributed Embedded Systems, Fuerstenallee 11, 33102 Paderborn, Germany,
e-mail: lisa@c-lab.de

Sebastian Korf
Heinz Nixdorf Institute, University of Paderborn, System and Circuit Technology, Fuerstenallee 11, 33102 Paderborn, Germany,
e-mail: korf@hni.uni-paderborn.de

Martin Krüger
Heinz Nixdorf Institute, University of Paderborn, Control Engineering and Mechatronics, Fuerstenallee 11, D-33102 Paderborn, Germany,
e-mail: kruemar@uni-paderborn.de

Tobias Meyer
Mechatronics and Dynamics, University of Paderborn, Pohlweg 47-49, 33098 Paderborn, Germany,
e-mail: tobias.meyer@uni-paderborn.de

Dr. Wolfgang Müller
Heinz Nixdorf Institute, University of Paderborn, Design of Distributed Embedded Systems, Fuerstenallee 11, 33102 Paderborn, Germany,
e-mail: wolfgang@c-lab.de

Jun.-Prof. Dr. Sina Ober-Blöbaum
Computational Dynamics and Optimal Control, University of Paderborn, Warburger Straße 100, 33098 Paderborn, Germany,
e-mail: sinaob@math.uni-paderborn.de

Dr.-Ing. Mario Porrmann
Cognitronics and Sensor Systems, Center of Excellence Cognitive Interaction Technology, Bielefeld University,Universitätsstraße 21-23, 33615 Bielefeld, Germany,
e-mail: mporrmann@cit-ec.uni-bielefeld.de

Claudia Priesterjahn
Heinz Nixdorf Institute, University of Paderborn, Software Engineering Group, Zukunftsmeile 1, 33102 Paderborn, Germany,
e-mail: c.priesterjahn@uni-paderborn.de

Prof. Dr. Franz-Josef Rammig
Heinz Nixdorf Institute, University Paderborn, Design of Distributed Embedded Systems, Fuerstenallee 11, 33102 Paderborn, Germany,
e-mail: franz@uni-paderborn.de

Christoph Rasche
Heinz Nixdorf Institute, University of Paderborn, Design of Distributed Embedded Systems, Fuerstenallee 11, 33102 Paderborn, Germany,
e-mail: crasche@c-lab.de

Peter Reinold
Heinz Nixdorf Institute, University of Paderborn, Control Engineering and Mechatronics, Fuerstenallee 11, 33102 Paderborn, Germany,
e-mail: peter.reinold@hni.upb.de

Prof. Dr. Wilhelm Schäfer
Heinz Nixdorf Institute, University of Paderborn, Software Engineering Group, Zukunftsmeile 1, 33102 Paderborn, Germany,
e-mail: wilhelm@uni-paderborn.de

Albert Seifried
University of Paderborn, Warburger Straße 100, 33098 Paderborn, Germany,
e-mail: seifried@math.uni-paderborn.de

Prof. Dr.-Ing. habil Walter Sextro
Mechatronics and Dynamics, University of Paderborn, Pohlweg 47-49, 33098 Paderborn, Germany,
e-mail: walter.sextro@uni-paderborn.de

Christoph Sondermann-Woelke
Mechatronics and Dynamics, University of Paderborn, Pohlweg 47-49, 33098 Paderborn, Germany,
e-mail: christoph.sondermann-woelke@uni-paderborn.de

Katharina Stahl
Heinz Nixdorf Institute, University of Paderborn, Design of Distributed Embedded Systems, Fuerstenallee 11, 33102 Paderborn, Germany,
e-mail: katharina.stahl@uni-paderborn.de

Dominik Steenken
Specification and Modelling of Software Systems, University of Paderborn, Warburger Straße 100, 33098 Paderborn, Germany,
e-mail: dominik@uni-paderborn.de

Robert Timmermann
Applied Mathematics, University of Paderborn, Warburger Straße 100, 33098 Paderborn, Germany,
e-mail: robert.timmermann@math.upb.de

Prof. Dr.-Ing. habil. Ansgar Trächtler
Heinz Nixdorf Institute, University of Paderborn, Control Engineering and Mechatronics, Fuerstenallee 11, 33102 Paderborn, Germany,
e-mail: ansgar.traechtler@hni.upb.de

Mareen Vaßholz
Heinz Nixdorf Institute, University of Paderborn, Product Engineering, Fuerstenallee 11, 33102 Paderborn, Germany,
e-mail: mareen.vassholz@hni.uni-paderborn.de

Prof. Dr. Heike Wehrheim
Specification and Modelling of Software Systems, University of Paderborn, Warburger Straße 100, 33098 Paderborn, Germany,
e-mail: wehrheim@uni-paderborn.de

Dr. Katrin Witting
Applied Mathematics, University of Paderborn, Warburger Straße 100, 33098 Paderborn, Germany,
e-mail: witting@math.upb.de

Tao Xie
Heinz Nixdorf Institute, University of Paderborn, Design of Distributed Embedded Systems, Fuerstenallee 11, 33102 Paderborn, Germany,
e-mail: tao@c-lab.de

Yuhong Zhao
Heinz Nixdorf Institute, University of Paderborn, Design of Distributed Embedded Systems, Fuerstenallee 11, 33102 Paderborn, Germany,
e-mail: zhao@uni-paderborn.de

Steffen Ziegert
Specification and Modelling of Software Systems, University of Paderborn, Warburger Straße 100, 33098 Paderborn, Germany,
e-mail: steffen.ziegert@uni-paderborn.de

Prof. Dr.-Ing. Detmar Zimmer
Design and Drive Technology, University of Paderborn, Pohlweg 47-49, 33098 Paderborn, Germany,
e-mail: detmar.zimmer@uni-paderborn.de

Chapter 1
Introduction to Self-optimization and Dependability

Abstract. This chapter gives an introduction to self-optimizing mechatronic systems and the risks and possibilities that arise with these. Self-optimizing mechatronic systems have capabilities that go far beyond those of traditional mechatronic systems. They are able to autonomously adapt their behavior and so react to outer influences, which can originate e.g. from the environment, changed user requirements or the current system status. The basic process of self-optimization, the procedures employed within and the main components of a self-optimizing system are explained here.

However, during the development of such a system, several challenges need to be met. On this note, both the concept of dependability and our proposed development process for self-optimizing systems are introduced. This process is used to derive a methodology for the selection of dependability methods in the development of self-optimizing systems.

To illustrate the proposed process and methods, several demonstrators are introduced. These are self-optimizing systems from different fields of engineering, e.g. robotics, automotive engineering and railroad engineering.

The chapter concludes with an overview of the content of the remainder of the book.

Tobias Meyer, Claudia Priesterjahn, and Walter Sextro

In today's market, a growing number of competitive and successful products are mechatronic systems. Mechatronic systems provide an advanced functionality that is achieved by a collaboration between mechanical engineering, electrical engineering, control engineering, software engineering, and mathematics. The combination of these domains allows the implementation of self-optimizing capabilities, which enable a system to react autonomously to changes in the system requirements or the environment. Moreover, high dependability of these systems is an absolute must, as self-optimizing mechatronic systems are often used in environments in which safety is critical. However, due to the complexity of self-optimizing mechatronic systems, ensuring dependability becomes a difficult task.

J. Gausemeier et al. (eds.), *Dependability of Self-optimizing Mechatronic Systems*,
Lecture Notes in Mechanical Engineering,
DOI: 10.1007/978-3-642-53742-4_1, © Springer-Verlag Berlin Heidelberg 2014

Fig. 1.1 Two RailCabs forming a convoy (artist's interpretation)

One practical example of a self-optimizing mechatronic system is the RailCab system, shown in Fig. 1.1. The vision of the RailCab project is a mechatronic rail system where small autonomous vehicles, called RailCabs, provide flexible transport on rail tracks. RailCabs form convoys without being rigidly coupled, in order to reduce energy consumption caused by air resistance. Such convoys are established on demand, and require only small distances between the RailCabs. The small distances required make the RailCabs a safety-critical system requiring high dependability.

No prior approach to the development of self-optimizing systems has taken the inherent risks and capabilities of these systems into account. The development process, which is presented in the book "Design Methodology for Intelligent Technical Systems" [14], coordinates the different domains that are involved in the development of a mechatronic system. However, this process needs to be complemented by methods enabling the developer to guarantee the dependability of self-optimizing mechatronic systems.

In order to overcome the apparent contradiction between the additional complexity introduced by self-optimization and the high dependability requirements, self-optimization capabilites within the system itself can be employed to increase the dependability of a system. This can be achieved by using behavior adaptation based on the current system state. In this book, we present methods for the development of self-optimizing mechatronic systems that either focus on ensuring the dependability of such systems or that use the capability of self-optimizing systems to increase their dependability system-internally. These methods are integrated into the development process for self-optimizing mechatronic systems in the book "Design Methodology for Intelligent Technical Systems" [14] and thus allow for early reaction to possible risks.

Using the development process presented in this book, the RailCab has become an example of a dependable self-optimizing mechatronic system.

1.1 Self-optimizing Mechatronic Systems

Michael Dellnitz, Kathrin Flaßkamp, Philip Hartmann, Martin Krüger, Tobias Meyer, Claudia Priesterjahn, Sina Ober-Blöbaum, Christoph Rasche, Walter Sextro, Katharina Stahl, and Ansgar Trächtler

Self-optimizing systems are intelligent mechatronic systems that are able to adapt their behavior autonomously to changing operating conditions and user demands. To this end, three steps are continuously cycled, as discussed in [12, 13], see also Fig. 1.2:

1. Analysis of current situation:
   During this step, the user requirements, the system state and based on direct observations the environment are identified and considered. Observations can also be indirectly obtained by communicating with other systems, and the state of the system may include prior observations. A major aspect of this step is the evaluation in how far the given objectives have been fulfilled.
2. Determination of objectives:
   In the second step, the new system objectives can be selected, adapted or generated from the system's previous objectives. Selection is possible if a finite number of discrete possible objectives exist. Adaptation is carried out if the system's objectives can be altered gradually. The generation of new objectives is performed if new objectives must be created independently of known objectives.
3. Adaptation of system behavior:
   The system behavior is adapted to account for changes necessitated by the determination of objectives. Changes to system parameters as well as changes to the system structure are possible. This action constitutes the feedback of the self-optimization cycle to the system.

As already mentioned, the objectives are of major importance for a self-optimizing system. They are organized in the System of Objectives, which in general is a set

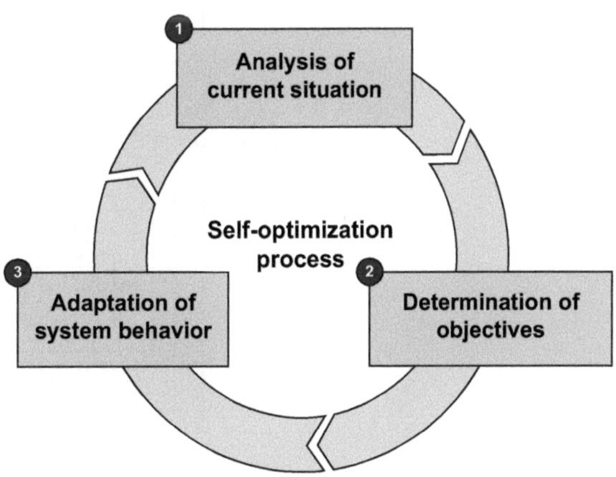

Fig. 1.2 Cycle of the behavior adaptation of a self-optimizing mechatronic system

of all objectives. If there are additional relations among the objectives, an objective hierarchy or an objective graph can be created.

Figure 1.3 shows the key aspects and the mode of operation of a self-optimizing system. Factors influencing the technical system originate in its surroundings (environment, users, etc.) or from the system itself, and can support the system's objectives or hinder them. Influences from the *environment*, for example strong winds or icy conditions, are unstructured and often unpredictable. If they hinder the system in fulfilling its pursued objectives, they are called disturbance variables. The *user* can influence the system, for instance by choosing preferred objectives. It is also possible for the *system* itself or other technical systems to influence the system's objectives, for example if mechanical components are damaged. Based on these changes in the environment, the *System of O bjectives* and therefore the system's *behavior* is adjusted. The adjustment of the system behavior can either be effected by changing control *parameters* or by changing the *structure*, e.g. through reconfiguration [13].

The objectives are the required, desired and/or undesired properties of the system. They are pre-determined during system development, but their desired characteristics are defined only during operation. Objectives are classified into external, internal and inherent objectives.

Inherent objectives constitute part of the functionality of the system, and are defined by the system itself. External objectives describe objectives that are characterized outside of the system; this can be due to other systems or human users, for example. Internal objectives are those objectives that are being pursued at any given moment by the system. They can be formulated using inherent and external objectives by selecting, weighting or setting them in relation to one another. Unless otherwise stated, the term objectives refers to internal objectives.

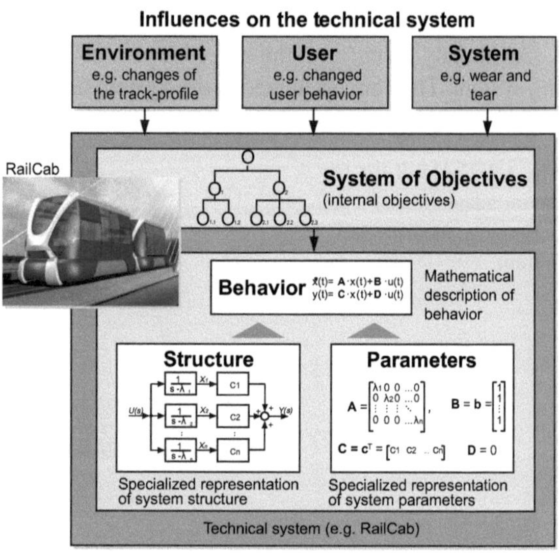

Fig. 1.3 Aspects of self-optimizing systems

Self-optimization, as described above, relates mainly to improving the performance of mechatronic system by considering the given or derived set of system objectives. Describing dependability attributes as system objectives allows engineers to use self-optimization as a means of achieving dependability. However, autonomous adaptations potentially introduce additional risks as well, because they may lead to unintended or even malicious system behavior. Additional mechanisms are required to assure the system's dependability at runtime in an autonomous manner. This may be accomplished by self-healing.

For the understanding of self-healing, we must first clarify what a *healthy* system is. Shaw [26] has defined a *healthy* system as follows: "A system is considered healthy as long as it works according to its specification." In other words, a system is *healthy* as long as it shows the intended behavior. *Self-healing* is the ability of a system to autonomously recover from an unhealthy system state. This means that in the case the system shows unintended behavior, it will recover and return to the desired behavior. Self-healing may consequently be used to prevent dangerous situations in the case of failure.

We can map the *self-healing process* onto the self-optimization process as described above. According to Gorla et al. [15], the self-healing process comprises the following steps: *failure detection*, *failure cause identification*, and *execution of self-healing operation*. The steps *failure detection* and *failure cause identification* are integrated into the step "Analysis of the current situation". The *execution of the self-healing operation* is part of the "Adaptation of the system behavior".

Failure detection requires knowledge of the system state, which is usually obtained by online monitoring. The system state is evaluated and classified as either a *healthy* or *unhealthy* state. Distinguishing between *healthy* and *unhealthy* system states is the most challenging part of failure detection. Runtime failure detection mechanisms may be implemented based on either system models, simulations results or anomaly detection methods [5].

When a failure has been detected, its cause needs to be located and its type determined [22] in order to select an appropriate self-healing operation in the next step. The failure can occur anywhere in the system, far away from the failure cause. The signal of a broken sensor, for example, may result in an actuator failure and only the actuator failure may be detected. The actuator and the sensor may be separated by many software components that process the sensor signal and output control commands. Moreover, the relation between failure cause and failure is usually not a one-to-one correlation. Failure causes may result in several failures, and/or a failure can be caused by many failure causes. The effect of a failure cause may result in a failure after only after long-term operation [15].

The self-healing operation removes the failure cause from the system and stops failures from propagating further. In this book, we particularly focus on self-healing in the form of structural reconfiguration. Once a failure cause has been identified, the failed components are removed from the basic system. This must happen before failures propagate to critical parts in the system, e.g., before they may cause hazards.

Self-healing is only applied if the system is in an unhealthy state. Failures are detected automatically and the system is returned to a healthy state without interrupting its operation. Self-healing thereby enhances the system's dependability by introducing robustness against failures.

1.1.1 Operator Controller Module (OCM)

The information processing unit of a self-optimizing system has to perform a multitude of functions: quasi-continuous control of the system motion, monitoring in view of occurring malfunctions, adaptation of the control strategy to react to changing environmental conditions and communication with other systems, to name just a few of these functions.

In order to ensure clear and manageable information processing, an architecture is needed which contains all these functions. The Operator Controller Module (OCM) [16] is an architecture with three levels that has been proved to be both an advantageous as well as an effective structure for self-optimizing systems. It consists of three levels, which can be also found in cognition science [27]. The architecture is shown in Fig. 1.4 and can be used for structuring the information processing of subsystems as well as of the entire system.

On the lowest level of an OCM, there is the *Controller*, which ensures the desired dynamical behavior of the plant and is similar to a classic control loop. Measurements are used to compute control signals for the actuators. The *Controller* operates in a quasi-continuous way under hard real-time constraints. Several types of controllers can be implemented at the same time with the possibility of switching between them.

The *Reflective Operator* monitors and regulates the Controller. It consists of sequential control, emergency routines and adaptation algorithms for the control strategies. The Reflective Operator does not access the actuators of the system directly, but modifies the Controller by initiating changes to the controller parameters or by switching between different controllers; this modification is undertaken by the Configuration Control. The *Reflective Operator* is event-oriented and has to operate under hard real-time constraints, because it is tightly linked to the Controller. However, it is also the connecting element to the cognitive level of the OCM and provides an interface between the Controller and those elements that are not capable to operate in real-time.

The topmost level of the OCM is represented by the *Cognitive Operator*. Various methods such as learning, use of knowledge-based systems or model-based optimization can be implemented on this level. The results can be used to improve the system behavior. This optimizing information processing can be roughly divided into model-based and behavior-based optimization, introduced in Sect. 1.1.2. While both the Controller and the Reflective Operator are subject to hard real-time constraints, the *Cognitive Operator* can also operate asynchronously to real time. Nevertheless, it has to respond within a certain amount of time. Otherwise, self-optimization would not find utilizable results in view of changing environmental conditions. Hence, the *Cognitive Operator* is subject to soft real-time.

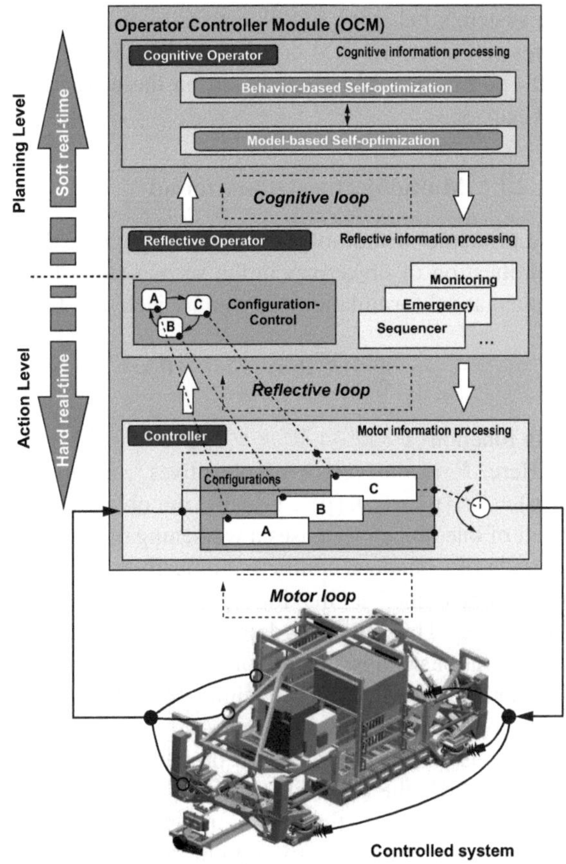

Fig. 1.4 Structure of Operator Controller Module

1.1.2 *Basic Procedures for Self-optimization*

To achieve the step "Determination of Objectives" of the self-optimization cycle, there are some basic concepts which we will introduce briefly in the following. Originally, these procedures were not necessarily intended to be self-optimization specific, but they turn out to be well suited for the behavioral adaptation of self-optimizing systems.

For self-optimizing systems, there typically arise a number of different objectives which have to be considered simultaneously. During operation, a self-optimizing system may, for instance, change from a primarily energy-saving or time-saving operation mode to another configuration which corresponds to a safer behavior or a longer lifetime. Firstly, this leads to multiobjective optimization problems (cf. Sect. 1.1.2.1), which have to be considered while still in the design phase of intelligent mechatronic systems. Secondly, during operation, Pareto optimal points, i.e. optimal compromises between conflicting objectives, have to be selected. This can be accomplished using planning techniques (cf. Sect. 1.1.2.2). To further adapt the system's behavior and objectives to environmental influences, learning procedures (cf. Sect. 1.1.2.3) can be applied. Specific methods for dependability and self-optimization which are based on these basic concepts are presented in detail in Chap. 3.

1.1.2.1 Multiobjective Optimization

The principle of multiobjective optimization enables the model-based, automated identification of objectives in the second step of the self-optimization process. A mathematical formulation of a multiobjective optimization problem[1] is given by

$$\min_{\mathbf{p}\in S}\{\mathbf{F}(\mathbf{p})\}, \quad S := \{\mathbf{p}\in\mathbb{R}^n | \mathbf{g}(\mathbf{p}) = 0, \mathbf{a}(\mathbf{p}) \leq 0\}, \tag{1.1}$$

with functions $\mathbf{F}:\mathbb{R}^n \to \mathbb{R}^k, k>1, \mathbf{g}:\mathbb{R}^n \to \mathbb{R}^m$ and $\mathbf{a}:\mathbb{R}^n \to \mathbb{R}^q$.

Here, $\mathbf{F}$ is the vector of objectives, i.e. functions $f_1,\ldots,f_k : \mathbb{R}^n \to \mathbb{R}$ with $\mathbf{F}(\mathbf{p}) = (f_1(\mathbf{p}),\ldots,f_k(\mathbf{p}))$. Usually, the objectives are conflicting, i.e. an improvement of one objective causes a worsening of the other objectives. Typical examples of trade-offs between conflicting objectives are "minimal energy consumption versus minimal time" or "maximal quality, but minimal time and minimal (e.g. production) costs". The objectives depend on the optimization parameters $\mathbf{p}$, e.g. controller configurations. An evaluation of the objective functions for a given set of optimization parameters often requires simulating a model of the continuous-time dynamical behavior of the technical system. Additionally, there may be equality or inequality conditions on the minimization problem, denoted by $\mathbf{g}$ and $\mathbf{a}$, respectively. They define $S \subset \mathbb{R}^n$, the admissible set of optimization parameters.

While ordinary optimization problems typically have a single global optimum, the solution of multiobjective optimization problems results in an entire set of Pareto points, the *Pareto set*. Minimization is intended with respect to the following partial ordering $\leq_p$ on $\mathbb{R}^n$: given $\mathbf{u},\mathbf{v}\in\mathbb{R}^n$, the vector $\mathbf{u}$ is smaller than the vector $\mathbf{v}$, $\mathbf{u} \leq_p \mathbf{v}$, if $u_i \leq v_i$ for all $i \in \{1,\ldots,k\}$. The solutions of multiobjective optimization problems can then be defined as follows: a point $\mathbf{p}^\star \in \mathbb{R}^n$ is called *globally Pareto optimal* for the multiobjective optimization problem (or a *global Pareto point*) if there does not exist any $\mathbf{p}\in\mathbb{R}^n$ with $\mathbf{F}(\mathbf{p}) \leq_p \mathbf{F}(\mathbf{p}^\star)$ and $f_j(\mathbf{p}) < f_j(\mathbf{p}^\star)$ for at least one $j \in \{1,\ldots,k\}$. That means, no other point $\mathbf{p}$ gives a better or equal (but not entirely identical) value in all objectives. However, there typically exist other Pareto optimal points which are, compared to $\mathbf{p}^\star$, better with regard to one objective but worse in another.

Figure 1.5 gives an illustration of a multiobjective optimization problem and Pareto optimal points. If the Pareto optimality property only holds for some neighborhood $U(\mathbf{p}^\star) \subset \mathbb{R}^n$, $\mathbf{p}^\star$ is called locally Pareto optimal. The image of the Pareto set, i.e. the corresponding function values, is called the *Pareto front* (cf. Fig. 1.5).

[1] By convention, only minimization problems are considered, since every maximization problem can be turned into a minimization problem.

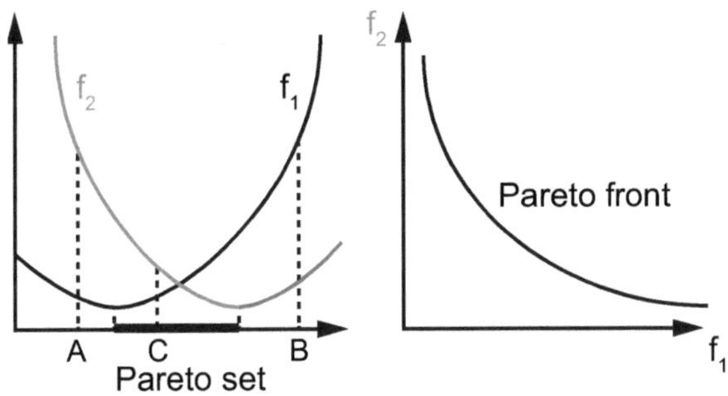

Fig. 1.5 Sketch of a multiobjective optimization problem with two objectives f_1 and f_2; while the points (A) and (B) are not optimal, (C) is a point of the Pareto set.

For the solution of real-world multiobjective optimization problems, numerical techniques have to be applied. There exist a number of methods for the computation of single Pareto points; for an overview we refer to [10]. However, for self-optimizing systems, it is important to gather knowledge about the entire Pareto set. In the last few decades, a number of techniques for the computation of entire Pareto sets have been developed (cf. e.g. [6, 7, 17, 21, 24]). In the course of the research of the Collaborative Research Center 614, set-oriented methods for multiobjective optimization (cf. e.g. [8] for an early reference or [25] for an overview) have been developed. Due to the approximation of the entire Pareto set (or the front, respectively) by box coverings, the methods are outstanding in their robustness and applicability to real-world multiobjective optimization problems, in particular for self-optimizing technical systems.

The computation of entire Pareto sets of multiobjective optimization or optimal control problems is computationally costly but important for the design of a knowledge base on which self-optimization relied during operation. The Pareto optimal alternatives are computed offline in advance and stored in this knowledge base. During operation, one specific optimal configuration of the Pareto set has to be chosen at any given time: this process is called *decision making*. Current information about the system's state (e.g. state of charge of the onboard power systems) and on the environment, such as weather conditions, can be taken into account in the decision-making process. To decide which Pareto optimal configuration should be applied, additional optimization or planning techniques can be used.

1.1.2.2 Planning

Any task for a mechatronic system (e.g. the transportation of persons or goods between two locations) can be expressed by a function [23]. This function defines the relationship between input and output variables [9] of the system by converting incoming energy, material and information flows into outgoing flows of the same types. Subtasks (such as driving with an active suspension system) are represented by analogous partial functions which are logically connected which make up the hierarchy of the overall function. The effect of a partial function also depends on the physical effect leading to a particular partial function solution [23]. Thus, partial functions can be implemented using various solutions (e.g. high or low compensation of disturbances). The choice of solutions to these partial functions determines the solution to the overall function and its effect on the mechatronic system.

Let PF_{leaf} be the set of all partial functions at the lowest level in the overall function hierarchy. Then, the selected solutions in the overall function at time t_j (see Fig. 1.6, black circles) can be listed as a sequence to solutions of partial functions [20]:

$$s_{of}(t) = (s_{pf_1}, ..., s_{pf_k}), \tag{1.2}$$

with $s_{pf_i} \in S_{pf_i}$ and $1 \leq i \leq k = |PF_{leaf}|$.

The behavior of the overall system at time t can be described by $V(t) = (\mathbf{x}_t, s_{of}(t), \mathbf{y}_t)$ with $\mathbf{x}_t$ as input vector and $\mathbf{y}_t$ as output vector. Consequently, the behavior of one time period is: $V(t_a, t_b) = (V(t_a), ..., V(t_b))$ with $t_a < t_b$ (see Fig. 1.6).

The mechatronic system reacts to each input variable by converting it to the output variables; it does so by executing the currently implemented overall function (linked partial functions of the functional hierarchy, reactive behavior). The information processing takes time and there is a latency involved according to the response to input variables. Because of that, the effect of the overall function on the output variables is delayed. Since the system execution takes place under real-time conditions, this has to be taken into account during the development of self-optimizing systems.

For autonomous control [9] of behavior, the system has to independently select solutions to the partial functions at certain times in order to achieve the desired effect (active behavior). The selection is made by considering specific objectives. This is the decision problem of aself-optimizing mechatronic system: selecting of solutions to partial functions which achieve these objectives with a high degree of fulfilment (goal-directed behavior). Mechatronic systems with advanced information processing are able to autonomously and proactively plan their behavior. The goal of this behavior planning is to minimize or maximize economic objectives, e.g. minimizing energy consumption. Consideration of the nondeterministic system environment and continuous processes are particular challenges. First, a planner is used to generate a plan for the behavior of the mechatronic system. Subsequently, the execution of this plan is carried out by the system.

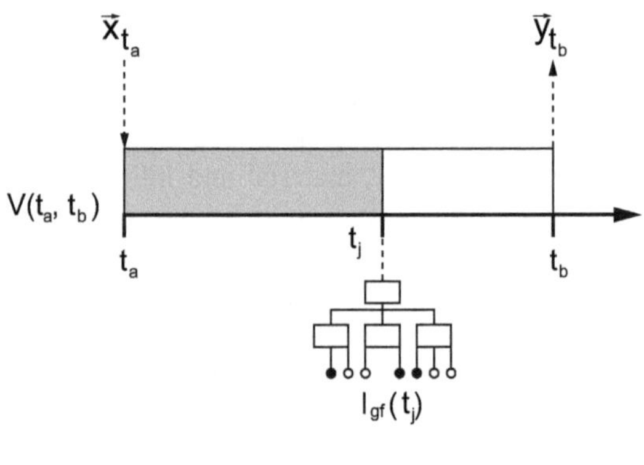

Fig. 1.6 Behavior of a mechatronic system

While planning is used to determine the optimal overall sequence of Pareto alternatives for task fulfillment (short term), learning is another approach to adapting the systems's behavior by itself based on data gathered by monitoring the environmental influences (long-term).

1.1.2.3 Learning

Apart from multiobjective optimization or planning several other approaches can be used to design a self-optimizing system. One of these approaches is machine learning. In general, machine learning involves the programming of computers to optimize a performance criterion based on example data or past experience [2]. Such an approach may be necessary when, for example, there is no human expertise available to solve a particular problem. Learning techniques may also be useful when a system's environment shows nondeterministic behavior or its behavior is not known.

Considering a complex mechatronic system moving from one destination to another destination, several unpredictable occurrences may crop up during movement. The system should be able to adapt itself as a reaction to the occurrences such that it can still optimize its performance criterion.

Another example would be if the problem to be optimized changes over time. If a mechatronic system plans its paths from one destination to another, this initial plan might at some point no longer be feasible, for instance due to traffic. In such a case, a learning program is able to adapt itself to the best path based on data gathered by monitoring traffic.

In learning approaches, a model is defined based upon certain parameters. Learning, then, is the optimization of the model parameters using training data or past experience. Using such an approach, it is possible to design a model capable of predicting the future [2].

Learning itself is a two-step procedure. The first step is called training. Efficient algorithms for solving the optimization problem itself, as well as for storing and processing the massive amount of data, are needed for the training phase. When the model has been trained, the second step, i. e. the use of this model during action, takes place. This step requires an efficient representation of the model and efficient algorithmic solutions for inference.

Machine learning is subdivided into three main learning techniques: supervised learning, unsupervised learning and reinforcement learning. The aim of supervised learning is to learn how to map input to an output, whereby the correct values are given by a supervisor. Such a supervisor is removed in unsupervised learning, which is based only on input data. The aim of unsupervised learning is to find regularities in the input. Normally, a certain kind of structure exists in the input data; thus, certain patterns occur more frequently than others and the system must determine what happens in general and what does not. In statistics, this is called density estimation, one example of which is clustering [2]. Another example is the learning of road disturbance profiles as described in Sect. 3.2.2. These road disturbances are described as a Bayesian network. Parameters and structure of this network are iteratively learned from observations.

The last learning technique introduced here is reinforcement learning. It is typically used to determine a sequence of actions such as planning; however, a planning approach, as described above, usually relies on a detailed description of a system's states and its actions, including their preconditions and effects, i.e. how they perform a transition from one state to another. Based on this knowledge, it determines a sequence of actions leading from an initial state to a goal state. If such knowledge is not available, reinforcement learning may be used to determine a sequence of actions to reach the goal. Such a sequence is called a policy. In this case, no single best action at every intermediate state exists. Thus, not a single action but rather a policy is important, and a single action is called good if it is part of a good policy [2]. The learning program should be able to assess whether a policy is good or not; in doing so, it then learns from good action sequences executed in the past and improves the policy with knowledge gained from these observations. This is often done using rewards, whereby the agent or program obtains a specific reward for each action. The final aim of the agent is to maximize some notion of the cumulative reward. Reinforcement learning is, for instance, used by the BeBot miniature robots for learning how to act in different environments: they adapt their behavior based on environmental changes. The approach is described in detail in *D.M.f.I.T.S*, [14], Sect. 5.3.9.

1.2 Challenges

Peter Iwanek, Tobias Meyer, Claudia Priesterjahn, Walter Sextro, and Mareen Vaßholz

Self-optimizing systems are usually subdivided into modules, each of which is then developed individually by a group of developers from all domains involved. All modules as well as all domains within one module have to be synchronized in order to result in a comprehensive system. This challenge can only be overcome with a development process and methods tailored to the tasks of self-optimizing systems. The proposed development process for self-optimizing systems is the core of the companion work to this volume, titled *Design Methodology for Intelligent Technical Systems* ("*D.M.f.I.T.S.*"), [14]. This companion work not only introduces the development process of self-optimizing systems, but also provides several methods which aid in creating the technology required for such systems. However, developing a self-optimizing system with its additional complexity brings along new risks concerning safety, reliability, maintainability and other attributes that are important for a dependable system. There is, of course, the possibility of improving some of these attributes by using the behavior adaptation capabilities of a self-optimizating system, thus counteracting the negative effects of self-optimization and, as a result, creating a self-optimizing system that offers better reliability, availability or is safer than a comparable conventional mechatronic system. These risks and possibilities are not covered in the book *D.M.f.I.T.S.* but need to be taken into account nonetheless.

1.2.1 Definition of the Term Dependability

The usage of the term dependability as it is used throughout this book is based upon the most common usage in computer sciences, which itself is well defined by Laprie et. al in [3]. According to this article, "*the dependability of a system is the ability to avoid service failures that are more frequent and more severe than is acceptable*", with a service failure being "*an event that occurs when the delivered service deviates from correct service*". This dependability is comprised of five *attributes*: availability, reliability, safety, integrity and maintainability[2].

In order to attain dependability in a system, several *means* are used. These can be grouped into four categories: fault prevention, fault tolerance, fault removal and fault forecasting. There are several *threats* to the dependability of a system because of failures, which are characterized by a deviation of at least one external state of the system from the correct service state, with the deviation called error and faults being the adjudged or hypothesized root cause. The complete taxonomy is depicted in Fig. 1.7.

To obtain a dependable self-optimizating system, all five attributes need to be considered. However, a self-optimizing system does not differ from a regular mechatronic system with regard to the attributes integrity and maintainability and thus can be treated much the same way; specialized methods are not required to make allowance for these factors. The three attributes that need to be regarded separately and for which prior methods are not fully suitable are availability, reliability and safety. Due to the interdisciplinarity of the development process, the means of achieving dependability are highly dependent on the development task.

Fig. 1.7 Dependability tree according to [3]

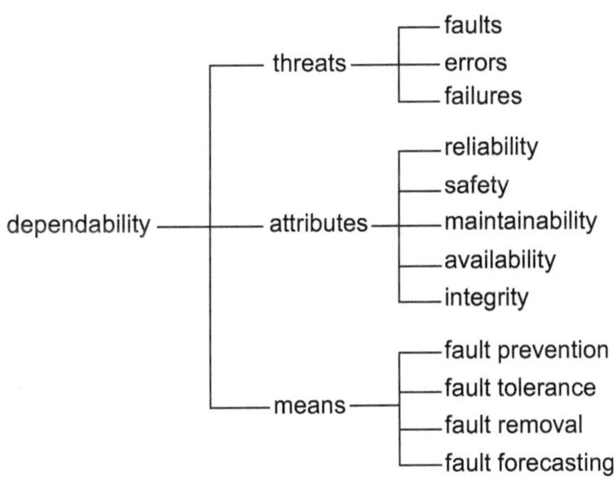

1.2.2 Dependability as a Challenge

Self-optimizing mechatronic systems are subject to additional threats to their dependability. This is not only due to their additional complexity, which increases the

[2] Note that confidentiality can be omitted, since it is not an attribute of dependability.

risk of systematic errors and makes the system more prone to failures, but also due to their non-deterministic behavior. The internal adaptation of the system makes it difficult to predict every system state in advance, making an exhaustive system analysis beforehand impossible. To obtain a safe system, special emphasis must be put on the software, which must be free of systematic flaws and which must be able to handle random faults. It must be possible to guarantee that the software meets the safety requirements.

In order to solve many engineering problems, domain-spanning or domain-specific models of the system are used. For example, in software development, models are used to guarantee the fulfillment of safety requirements and to identify systematic faults via verification. Random faults are analyzed using model-based hazard analysis. The program source code is then generated automatically from the system models.

However, any model can only represent a limited excerpt of reality. To handle the complexity of the models, it is necessary to omit certain aspects during modeling and thus to idealize a given real system. For example, uncertain parameters or perturbations can only be modeled with a limited level of detail. All these limitations lead to a deviation between the real system and the model. If a model is used to synthesize further elements of the system, e.g. if a model of the system dynamics is used to design a controller, such deviations can lead to unexpected problems when exchanging the model for the real system. While it is certainly possible to model uncertainties, only known uncertainties and anticipated magnitudes can be considered.

To overcome these limitations, detected deviations or uncertainties can be compensated for during system operation and potential failures or failures that have been detected during operation can be counteracted by the use of self-optimization, thus increasing the system's dependability. Yet while exploiting self-optimization as a means of increasing the dependability is a real possibility, the additional threats that arise with it must be controlled at the same time.

By introducing the use of additional methods during the development process, these aspects can be taken into account. However, this is not sufficient if the methods are used inefficiently or without adequate system data as input. Additionally, it might not be easily possible to select suitable methods in the first place.

A self-optimizing system relies heavily on the close integration of mechanical, electrical and software components, making the development not only of the system itself, but also of the system's dependability, a joint effort of developers from several domains. This requires domain-spanning methods, making collaboration between developers even more challenging.

1.2.3 Ensuring Dependability during the Development Process

The development of dependable self-optimizing systems is challenging due to the multidisciplinarity of mechatronics, the cross-linking between the subsystems, the lack of current knowledge in the fields of advanced mathematics and artificial intelligence, and increased dependability requirements. Existing and established approaches for the development of mechatronic systems such as the VDI guideline 2206 [28], the approach by Isermann [18] or Ehrlenspiel [11], the iPeM-Modell [1] or the V-Model by Bender [4], do not address self-optimization specifically and therefore cannot fulfill the requirements set by a self-optimizing system. To overcome these shortcomings, a design methodology[3] for self-optimizing systems was developed by the Collaborative Research Center 614. It expands the existing methodologies and supports developers by providing domain-spanning and self-optimization-specific methods and tools [19]. The methodology consists of a reference process, methods and tools for the development of self-optimizing systems. It is presented briefly in Chap. 2 and in detail in the book "Design Methodology for Intelligent Technical Systems" *D.M.f.I.T.S*, [14], Chap. 3.

The design methodology for self-optimizing systems supports the developers with knowledge and methods from self-optimization experts, offering an expertise which is generally not available in companies. The reference process provides recommendations for how to apply self-optimization methods, and therefore complements the existing development processes. For the application of the reference process to a specific development task and enterprise, the development process has to be applied on a case-by-case basis. To ensure the dependability of self-optimizing systems, te current book extends the presented methodology with dependability-specific methods. In practice, the developers are confronted with a large number of methods increasing the dependability of a system. To ensure the dependability of the system, it is necessary to select those engineering methods that are suitable to the underlying development task. This selection is tedious and error-prone; thus, the approach presented in Sect. 3.3 supports the developer during the selection of the appropriate dependability engineering method.

The methodology for improving the dependability of self-optimizing systems consists of a method database, a guide to the selection and planning of dependability engineering methods, and a software tool. The methods of the database that are appropriate for ensuring the dependability of a self-optimizing system are presented in Chap. 3. These methods need to be implemented during certain development phases. The procedure for using the method is integrated into the reference process for each selected dependability-specific method. Moreover, the interactions between the methods are considered, because some methods need the output of another as input. Therefore, both methods need to be employed during the development to get the desired result. However, even applying the methodology for improving dependability of self-optimizing systems,it cannot be guaranteed that a fully dependable system results. As with most engineering problems, certain assumptions and deliberate choices have to be made, which may all result in possible failures or in critical system states being neglected. In summary, the presented methodology can reduce the risk of failures during the development process, but nevertheless care must be taken to not omit important aspects and to always use proper engineering judgment.

[3] In this book the following definition of the used terms method and methodology provided by Pahl et al. are used: a method explains a procedure to reach a certain target; a methodology explains a procedure including several methods and tools [23].

1.3 Applications of Self-optimizing Systems

Christian Hölscher, Jan Henning Keßler, Tobias Meyer, Christoph Rasche, Peter Reinold, Walter Sextro, Christoph Sondermann-Woelke, and Detmar Zimmer

Several self-optimizing systems have been developed using the processes and methods explained within this book. Throughout this book, these systems are used as successful examples of how to apply these methods. These systems all have in common that they closely integrate mechanics, electronics and information technology. They use real-time hardware to run sophisticated control algorithms which produce the desired system behavior by means of electrical or hydraulic actuators. There are several contradictory objectives that need to be pursued at the same time. Depending on the current situation, these objectives are selected or characterized, thus forming the system behavior. On one hand, the challenge is the sheer complexity of these systems, with additional challenges arising when several systems are collaborating, e.g. RailCabs that have formed a convoy. On the other hand, the major possibility which opens up is an increase of the system's reliability and availability by adapting the system behavior when existing or prospective faults are detected.

As it is shown later on, all these challenges can be overcome and chances taken advantage of using a suitable development process with supplemental methods, as described in the following chapters.

1.3.1 Rail Technology – The RailCab System

The RailCab is an innovative rail-bound transportation system. It consists of autonomous vehicles, so-called RailCabs, which are capable of driving individually on regular rail tracks. In order to save energy, they can form convoys without being mechanically coupled by following one another at small distances. They are equipped with an active guidance system which enables a single RailCab to autonomously leave convoys, e.g. at passive switches.

Several safety risks for both the vehicles themselves and for passengers are associated with the forming of convoys and the passive switches used. To allow safe transportation, these had to be reduced and eliminated.

In order to simplify system development, the RailCab vehicles consist of separate modules for different basic functions. In the following chapters, several of these modules are used as example systems. Special attention is given to the intelligent Drive Module (iDM), which is required for propulsion by a linear drive unit, the Active Guidance Module as a steering subsystem, and the Active Suspension Module, which actively compensates for undesired body motions.

1.3.1.1 Intelligent Drive Module

The Driving efficiency of the RailCab depends on the distance, i.e. the air gap, between the two linear motor parts, called the primary- and the secondary part. The intelligent Drive Module has the task of minimizing the air gap in order to

Fig. 1.8 Test rig of the intelligent Drive Module

increase the efficiency, while ensuring collision-free operation at the same time. The minimization depends on the RailCab velocity and the track characteristics in relation to the dynamic behavior of the adjustment actuator.

The iDM test rig, see Fig. 1.8, consists of two acuator groups. Each of them is comprised of the primary part of a linear drive and an air gap adjustment actuator. Additionally, there are spring assemblies to reduce the actuator load and to raise the primary part in case of an actuator failure. Hall sensors on each actuator group measure the magnetic field of the secondary parts to determine the electric rotation angle, the track velocity and the air gap. The secondary parts of the iDM are arranged in a circle and can be adjusted vertically to simulate different track characteristics (see Fig. 1.9). In contrast to the RailCab, a synchronous motor without the capability of an energy transfer is installed in the test rig, because the focus of

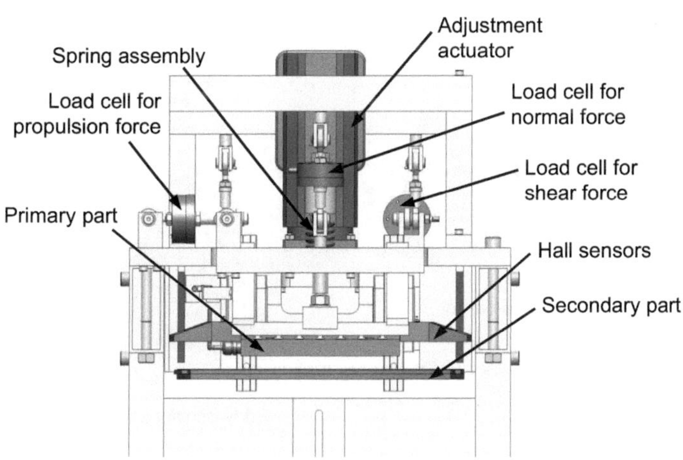

Fig. 1.9 Actuator group of the intelligent Drive Module

current work is on the improvement of the driving efficiency via an active air gap adjustment.

If the air gap is too small, a collision between the primary and secondary parts occur. Since this could lead to catastrophic failures of the whole system, the information processing not only has to be set up to compensate for such failures, but the mechanical system itself must also be able to cope with sudden actuator failures.

1.3.1.2 Active Guidance Module

At the high velocities of the RailCabs (intended maximum velocity is approximately 180 km/h), normal switches would not be able to change direction sufficiently quickly to be able to split up convoys; hence, passive switches are used. When passing over these, each RailCab steers individually towards its desired direction. The system module for this steering action is called the Active Guidance Module and is shown in Figure 1.10. In addition to steering on passive switches, the guidance module actively controls vehicle guidance on normal tracks. This reduces wear on both the wheels and the rails because flange contacts on straight tracks as well as in curves are avoided. Consequently, the rigid coupling of both wheels, as within normal rail wheel sets, is omitted and wheel slip is minimized. Further, disturbances such as track irregularities and cross wind are compensated for. There are also other influences on the Active Guidance Module that originate from the RailCab itself, such as, for example, the necessity of restricting energy consumption or varying the velocity. For the steering actions on the track, an optimal trajectory is calculated, which is based on the current and stored values of the clearance as well as on the lateral displacement. Both values are measured by installed redundant eddy-current sensors.

However, if the steering fails, the direction might not be deterministically selectable, thus making the RailCab prone to derailments. To eliminate this risk,

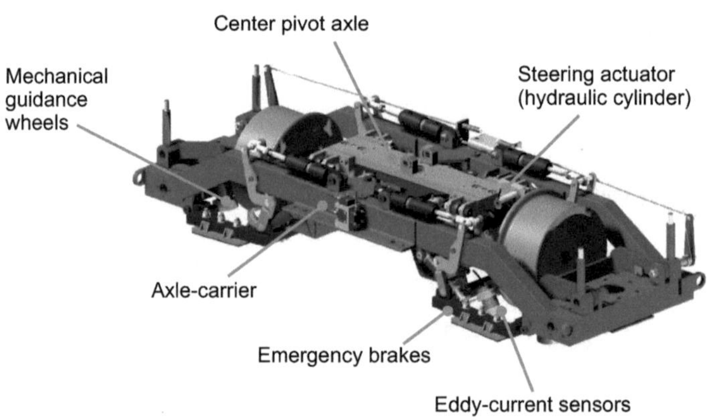

Fig. 1.10 Active Guidance Module of the RailCab

Fig. 1.11 Test rig of the Active Suspension Module

several counter-measures, in both hardware and information processing, have been incorporated.

1.3.1.3 Active Suspension Module

The Active Suspension Module of the RailCab is designed to increase passenger comfort. The test rig of the Active Suspension Module is illustrated in Fig. 1.11 and its hierarchical structure is shown in Fig. 1.12. The main task of this module is a nearly complete decoupling of the coach body from the excitations of the railroad track, especially in the upper frequency range. Its secondary task is to tilt the coach body into curves in order to reduce the lateral acceleration for passengers. Both can be achieved by a secondary suspension mounted under the chassis floor. The

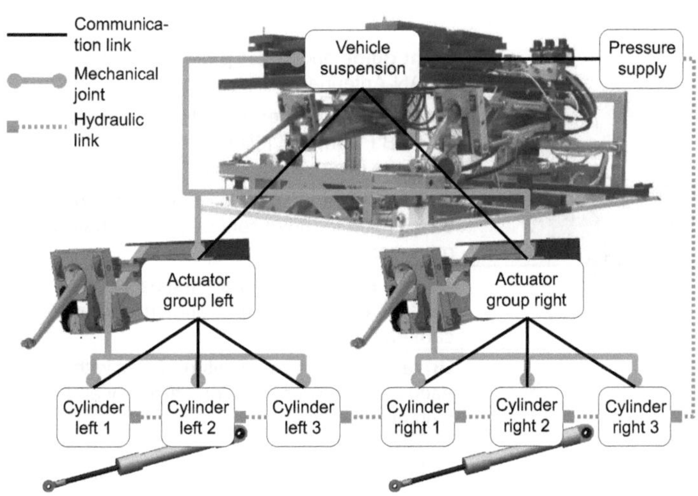

Fig. 1.12 Structure of the Active Suspension Module

active suspension of one RailCab consists of four actuator groups, each with a passive spring made of glass-fiber reinforced plastic (GRP) with a low damping characteristic and three hydraulic cylinders. By transmitting the motion of the cylinders via sophisticated guide kinematics to the spring base, the motion of the coach body can be damped and in some cases strong vibrations can be absorbed almost completely.

A failure of one such module would not only lead to a less comfortable ride, but also to possibly dangerous coach body excitations which would necessitate a lowered top speed. Thus, intelligently exploited redundancy for the actuators at risk has to be implemented.

1.3.2 Miniature Robot BeBot

The BeBot, illustrated in Fig. 1.13, is a small robot serving as a technology platform. It is used for research in the domains of dynamic reconfigurable systems, multi-agent systems and swarm intelligence. The dimensions of the miniature robot BeBot are approximately 9cm $\times$ 9cm $\times$ 9cm. The chassis is a molded interconnect device (MID). An embedded Linux operating system allows the execution of programs directly on the robot. The robot is equipped with a camera and infrared distance sensors to observe its environment. Two chain drives provide the robot with robust motion capabilities, even on rough ground. Furthermore, the BeBot is equipped with a light guide and can be extended with different modules to interact with its environment. Further information can be found in *D.M.f.I.T.S*, [14], Sect. 2.2.

The BeBots are used as a system to test various self-optimizing applications, one example of which would be dynamic threshold adaption. A Bebot activates or deactivates energy efficiency measures depending on the state of its batteries. The

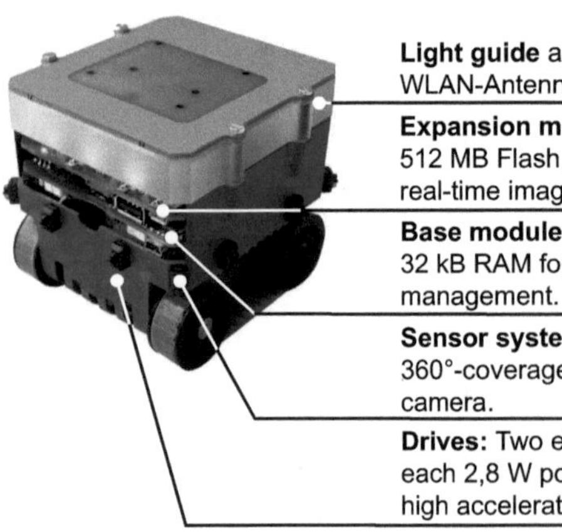

Fig. 1.13 The miniature robot BeBot: the two boards of the robot, the sensor system, the light guide and the chain drive are depicted

thresholds for its decisions about the activation or deactivation change dynamically due to the battery capacity. This capacity decreases over the course of its life period.

Another example is the use of learning strategies for statistical planning approaches, combined with partially observable Markov decision process (POMDPs). In this scenario, the BeBots learn to act in different environments. They adapt their behavior based on environmental changes. The statistical planning approach is described in detail in *D.M.f.I.T.S*, [14], Sect. 5.3.9, and includes the previously mentioned dynamic threshold adaptation.

1.3.3 X-by-Wire Test Vehicle

The demonstrator "Chameleon" is an X-by-wire vehicle, meaning that there are no mechanical couplings between the control elements and the actuators; it is shown in Fig. 1.14. This test vehicle is entirely actuated electrically and has four corner modules with three DC motors each: one for steering, one for driving and one for an active suspension. These enable an all-wheel drive, a single-wheel steering and an active suspension. The driver is able to control the vehicle dynamics via a joystick. The maximum speed is about 50 km/h and the empty weight is appr. 280 kg.

The driving motors not only drive but also brake, i.e. decelerate the vehicle. Additionally, the single-wheel steering enables the deceleration of the vehicle by steering the wheels inwards. With these two possibilities and their combination, it is up to the information processing to decide which possibility is optimal for the current situation, also considering wear and possible breakdown of actuators.

Fig. 1.14 Test vehicle "Chameleon"

1.4 Structure of This Book

Tobias Meyer and Walter Sextro

After giving a short outline of the possibilities and challenges when developing dependable self-optimizing mechatronic systems, the necessity of a concise, systematic approach has been made apparent. Our suggested approach is presented in the following chapters, an overview of which is given in this section.

In Chap. 2, a detailed introduction to the development process of self-optimizing mechatronic systems is given. This introduction is focused on dependability-related steps, which can be augmented by suitable methods.

Chapter 3 explains these methods in detail and also gives individual examples. It is meant to be used as a reference, with all methods having the exact same basic outline and with both required prerequisites and the output being mentioned prominently. Its main sections 3.1 and 3.2 are structured according to the phases of the development process.

Section 3.3 explains the process of selecting methods to be conducted during the development process. For this purpose, selection guidelines are introduced.

One example spanning the complete development process is given in Chap. 4. In this, the selection of methods using the guidelines from Sect. 3.3 is explained and the individual method's input and results are mentioned; the primary focus is on the interaction of several methods that are employed consecutively.

The book concludes with Chap. 5, which contains a summary of the results. Additionally, suggestions for further research are made.

References

1. Albers, A.: Five Hypotheses about Engineering Processes and their Consequences. In: Proceedings of the 8th International Symposium on Tools and Methods of Competitive, Ancona, IT (2010)
2. Alpaydın, E.: Introduction to Machine Learning. The MIT Press (2004)
3. Avizienis, A., Laprie, J.C., Randell, B., Landwehr, C.: Basic Concepts and Taxonomy of Dependable and Secure Computing. IEEE Transactions on Dependable and Secure Computing 1(1), 11–33 (2004), doi:10.1109/TDSC.2004.2
4. Bender, K.: Embedded Systems – Qualitätsorientierte Entwicklung. Springer, Berlin (2005), doi:10.1007/b138984
5. Chandola, V., Banerjee, A., Kumar, V.: Anomaly detection: A survey. ACM Comput. Surv. (2009), doi:10.1145/1541880.1541882
6. Coello Coello, C.A., Lamont, G., van Veldhuizen, D.: Evolutionary Algorithms for Solving Multi-Objective Optimization Problems, 2nd edn. Springer, Berlin (2007), doi:10.1007/978-0-387-36797-2
7. Das, I., Dennis, J.: A Closer Look at Drawbacks of Minimizing Weighted Sums of Objectives for Pareto Set Generation in Multicriteria Optimization Problems. Structural Optimization 14(1), 63–69 (1997), doi:10.1007/BF01197559

8. Dellnitz, M., Schütze, O., Hestermeyer, T.: Covering Pareto Sets by Multilevel Subdivision Techniques. Journal of Optimization Theory and Application 124(1), 113–136 (2005), doi:10.1007/s10957-004-6468-7
9. Deutsches Institut für Normung e.V.: DIN 19 226 Teil 1: Leittechnik – Regelungstechnik und Steuerungstechnik – Allgemeine Grundbegriffe. German National Standard (1994)
10. Ehrgott, M.: Multicriteria Optimization, 2nd edn. Springer, Berlin (2005), doi:10.1007/3-540-27659-9
11. Ehrlenspiel, K.: Integrierte Produktentwicklung, 3rd edn. Carl Hanser Verlag, Munich (2007)
12. Gausemeier, J. (ed.): Selbstoptimierende Systeme des Maschinenbaus, *HNI-Verlagsschriftenreihe*, vol. 155. Heinz Nixdorf Institute, University of Paderborn, Paderborn, DE (2004)
13. Gausemeier, J., Rammig, F.J., Schäfer, W. (eds.): Selbstoptimierende Systeme des Maschinenbaus. HNI-Verlagsschriftenreihe, vol. 234. Heinz Nixdorf Institute, University of Paderborn, Paderborn, DE (2009)
14. Gausemeier, J., Rammig, F.J., Schäfer, W. (eds.): Design Methodology for Intelligent Technical Systems. Lecture Notes in Mechanical Engineering. Springer, Heidelberg (2014), doi:10.1007/978-3-642-45435-6_2
15. Gorla, A., Mariani, L., Pastore, F., Pezzè, M., Wuttke, J.: Achieving Cost-Effective Software Reliability Through Self-Healing. Computing and Informatics 29(1), 93–115 (2010)
16. Hestermeyer, T., Oberschelp, O., Giese, H.: Structured Information Processing for Self-Optimizing Mechatronic Systems. In: 1st International Conference on Informatics in Control, Automation and Robotics, Setubal, PT (2004)
17. Hillermeier, C.: Nonlinear Multiobjective Optimization – A Generalized Homotopy Approach. Birkhäuser, Berlin (2001)
18. Isermann, R.: Mechatronische Systeme – Grundlagen. Springer, Berlin (2008), doi:10.1007/978-3-540-32512-3
19. Kahl, S., Gausemeier, J., Dumitrescu, R.: Interactive Visualization of Development Processes. In: Proceedings of the 1st International Conference on Modelling and Management of Engineering Processes (2010)
20. Klöpper, B.: Ein Beitrag zur Verhaltensplanung für interagierende intelligente mechatronische Systeme in nicht-deterministischen Umgebungen, HNI-Verlagsschriftenreihe, vol. 253. Heinz Nixdorf Institute, University of Paderborn, Paderborn, DE (2009)
21. Knowles, J., Corne, D., Deb, K.: Multiobjective Problem Solving from Nature: From Concepts to Applications. Springer, Heidelberg, doi:10.1007/978-3-540-72964-8
22. Lanigan, P.E., Kavulya, S., Narasimhan, P., Fuhrman, T.E., Salman, M.A.: Diagnosis in Automotive Systems: A Survey. Tech. Rep. CMU-PDL-11-110, Carnegie Mellon University Parallel Data Lab (2011), doi:10.1.1.208.1611
23. Pahl, G., Beitz, W., Feldhusen, J., Grote, K.H.: Kontruktionslehre – Grundlagen erfolgreicher Produktentwicklung – Methoden und Anwendung, 6th edn. Springer, Heidelberg (2005), doi:10.1007/b137606
24. Schäffler, S., Schultz, R., Weinzierl, K.: A Stochastic Method for the Solution of Unconstrained Vector Optimization Problems. Journal of Optimization Theory and Applications 114(1), 209–222 (2002), doi:10.1023/A:1015472306888
25. Schütze, O., Witting, K., Ober-Blöbaum, S., Dellnitz, M.: Set Oriented Methods for the Numerical Treatment of Multi-Objective Optimization Problems. In: Tantar, E., Tantar, A.-A., Bouvry, P., Del Moral, P., Legrand, P., Coello Coello, C.A., Schütze, O. (eds.) EVOLVE- A Bridge between Probability. SCI, vol. 447, pp. 187–219. Springer, Heidelberg (2013)

26. Shaw, M.: "Self-healing": softening precision to avoid brittleness: position paper for WOSS 2002: workshop on self-healing systems. In: Proceedings of the First Workshop on Self-Healing Systems, WOSS 2002, pp. 111–114. ACM, New York (2002), doi:10.1145/582128.582152
27. Strube, G.: Modelling Motivation and Action Control in Cognitive Systems. In: Mind Modelling, pp. 89–108. Pabst, Berlin (1998), doi:10.1.1.89.9346
28. Verein Deutscher Ingenieure (VDI): VDI 2206 – Entwicklungsmethodik für mechatronische Systeme. Technical Guideline (2004)

Chapter 2
Development of Self-optimizing Systems

Abstract. In the development of self-optimizing systems, developers have to face different challenges, such as the multidisciplinarity of mechatronics, cross-linking between the subsystems, the lack of current knowledge in the fields of advanced mathematics and artificial intelligence, and increased dependability requirements. To support the developer in this challenging task adequately, the Collaborative Research Center 614 has developed a design methodology consisting of a reference process, methods and tools. The reference process is divided into two phases: the "Domain-Spanning Conceptual Design" and the "Domain-Specific Design and Development". In the first phase, the domain-spanning model-based aproach CONSENS (CONceptual design Specification technique for the ENgineering of complex Systems) is used to create a common understanding basis between the different domains involved. Based on the Principle Solution developed in this phase, the "Domain-Specific Design and Development" is planned and implemented. To ensure the development of dependable self-optimizing systems, domain-spanning and domain-specific dependability engineering methods can be used. The developer encounters a significant number of these methods, that have to be integrated into the process to obtain an individualized development process for the specific development task.

Jürgen Gausemeier and Mareen Vaßholz

The Collaborative Research Center 614 has developed a design methodology that expands existing ones for mechatronic systems such as the VDI guideline 2206 [12], the approach by Isermann (2008) [8] or Ehrlenspiel (2007) [4], the iPeM-Modell [1] or the V-Model by Bender (2005) [2] and supports developers by providing domain-spanning and self-optimization-specific methods and tools [9]. The methodology consists of a reference process, methods and tools for the development of self-optimizing systems, and is presented in detail in the book "Design Methodology for Intelligent Technical Systems" *D.M.f.I.T.S*, [7], Chap. 3.

During the development of self-optimizing systems, different domains, such as mechanical, electrical/electronic, software and control engineering and experts from

J. Gausemeier et al. (eds.), *Dependability of Self-optimizing Mechatronic Systems*,
Lecture Notes in Mechanical Engineering,
DOI: 10.1007/978-3-642-53742-4_2, © Springer-Verlag Berlin Heidelberg 2014

advanced mathematics and artificial intelligence, are involved. Therefore, a domain-spanning development process is needed which supports common understanding of the product development activities and their cross-domain relationship for all participating developers. The development of self-optimizing systems (cf. Fig. 2.1) is basically structured into two main phases: the "Domain-spanning Conceptual Design" and the "Domain-specific Design and Development". The result of the first phase is the specification of a Principle Solution for the system to be developed. It describes the basic structure and operational mode of the system, as well as its desired behavior, computer-internally as partial models. In particular, the specification of the Principle Solution contains the description of the System of Objectives (cf. Sect. 2.1). Based on the Principle Solution, a first analysis of the dependability can be carried out.

This specification of the Principle Solution constitutes the basis for the Domain-specific Design and Development (cf. Sect. 2.2). During this phase, the domains involved work in parallel and use their domain-specific methods, models and tools. The partial solutions developed by the domains involved are continuously integrated into the overall solution with model transformation and synchronization techniques [11]. During the Domain-specific Design and Development, the ability of the system to adapt its behavior during operation is implemented in the domains. For example, multiobjective optimization is used in the domain control engineering to enable the system to adapt its behavior by changing the control parameter during operation.

These steps particularly require the involvement of experts of advanced mathematics and artificial intelligence. The optimization is specific to self-optimizing systems, as the main focus is on the implementation of the self-optimization process. In addition, system tests are performed; both virtual and real prototypes are used for

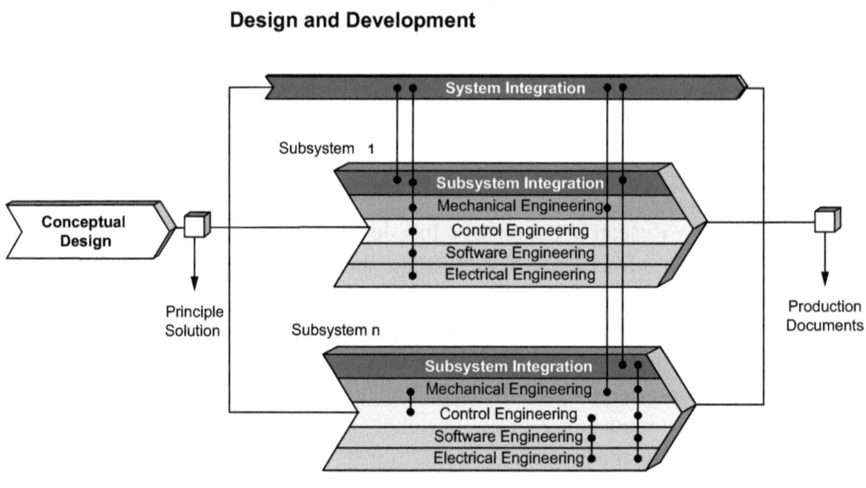

Fig. 2.1 Macro cycle for the development of self-optimizing systems

this purpose. The result of the Design and Development phase is engineering data such as manufacturing drawings and part lists.

The reference process for the development of self-optimizing systems is recommendatory and gives an overview of the tasks that have to be performed during the development. It is based on our experiences from the development of the RailCab and its function modules. It serves as basis for the application to a specific development task and company in form of an implementation model. This model provides a detailed process sequence according to the project, the system type and the environment of the development project. It consists of process steps from the reference process and builds the starting point for the project management. For example the implementation model for the RailCab consists of about 850 steps and 900 development objects. During the development project execution, deviations and changes from the planned development process can occur, for example due to changing development objectives such as time and costs. To support the management during the project execution to react to these changes we adopted the self-optimization approach to the management of the development process (cf. *D.M.f.I. T.S*, [7], Sect. 3.5). In the following sections the reference process is described in more detail.

2.1 Domain-Spanning Conceptual Design

The aim of the "Domain-spanning Conceptual Design" is to create a common understanding of the system between the domains involved. The main result of this phase is the Principle Solution, which is described by the specification technique CONSENS.

There are eight separate but highly interrelated aspects that need to be covered by the Principle Solution. Computer-internally, these aspects are represented as *partial models*. The eight aspects and their respective partial models are shown in Fig. 2.2 (cf. [6]). While each partial model represents only one individual aspect of the system, together they form a coherent model of the complete system [6]. The partial models are:

Environment:
: This model describes the environment of the system to be developed and the system's embedment into the environment. Relevant spheres of influence and influences will be identified and interactions between them will be examined [6].

Application Scenarios:
: These scenarios concretize the behavior of the system in a particular state and situation. Application Scenarios characterize a problem which needs to be solved in certain cases and also briefly outline the possible solution [6].

Requirements:
: This partial model presents an organized collection of requirements that need to be fulfilled during product development. Every requirement is described textually and, if possible, concretized by attributes and their characteristics [6].

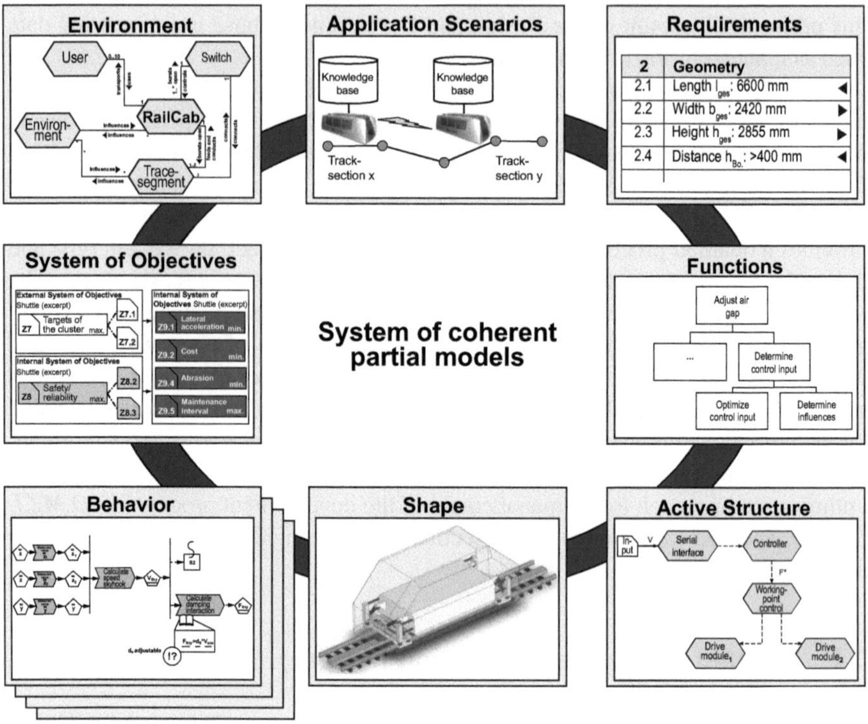

Fig. 2.2 Partial models of the domain-spanning description of the Principle Solution of a self-optimizing system [6]

Functions:

A function is the general and required relationship between input and output parameters with the aim to fulfill a task. Functions are implemented using solution patterns and their concretizations. Functions should be specified in a hierarchical manner [6].

Active Structure:

This partial model describes the system elements (e.g. drive and brake modules, energy management) and their attributes, as well as their relations (energy, material or information flow). Incoming parameters are also described (e.g. comfort) [6].

Shape:

The Shape is usually created by a 3D CAD-System and gives a first impression of the physical appearance of the system [6].

Behavior:

The partial model Behavior is subdivided into the aspects Behavior – State and Behavior – Activities. The aspect Behavior–State defines the states of the system and the state transitions. The state transitions describe the reactive behavior of the system towards incoming events, which can also be system activities. The aspect Behavior – Activities describes the logical sequence of activities which are carried out by the different system elements. Activities describe how functions are executed within different system states [6].

System of Objectives:
This partial model includes external, inherent and internal objectives, as well as the connections between them. Internal objectives are achieved by the system itself during operation mode and as a consequence represent the system's intentionality [6] (see also Sect. 1.1).

In particular, the partial model System of Objectives is of high importance for self-optimizing systems: it describes the objectives of the system during operations, their hierarchical relationship, and potential conflicts between objectives [10]. The specification of the Principle Sution provides a holistic domain-spanning description of the self-optimizing system and forms the basis for the communication and cooperation between the developers from the domains involved. As such, it enables them to avoid design mistakes and corresponding iteration loops in later development phases, which would otherwise occur due to differences in understanding of the system [3].

To develop the Principle Solution the reference process for the Domain-spanning Conceptual Design consists of four main phases: "Planning and Clarifying the Task", "Conceptual Design on the System Level", "Conceptual Design on the Subsystem Level" and "Concept Integration" (cf. Fig. 2.3).

In "Planning and Clarifying the Task", the design task of the system and the requirements are identified and evaluated. The results are the List of Requirements, the Environment model, the recommended product structure type and the design rules for this, as well as the Application Scenarios.

Based on previously determined requirements the system must fulfill, the main functions of the system are identified and set into a function hierarchy in the "Conceptual Design on the System Level". Each function has to be fulfilled to satisfy the requirements; therefore solution patterns are sought which can execute the desired functions. Within a morphologic box, the solution patterns are combined to obtain consistent solution variants. These are evaluated and the best one is chosen and integrated into the Principle Solution on the system level. The resulting solution does

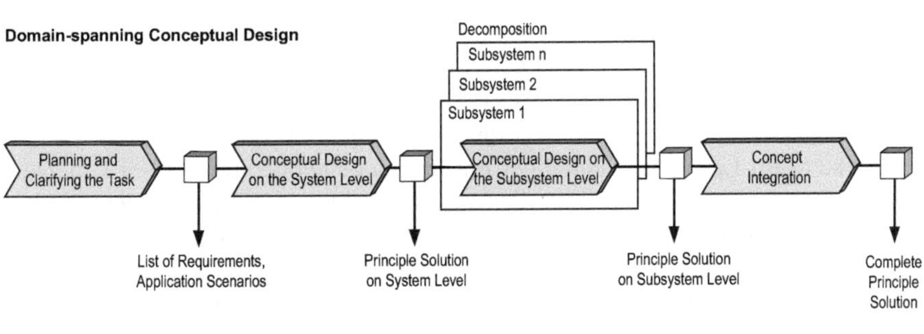

Fig. 2.3 The four main phases of the Domain-spanning Conceptual Design

not have to be self-optimizing at this stage. The consistent grouping of solution patterns allow the modeling of the Active Structure, an initial construction shape and the System Behavior.

At this point the potential for the use of self-optimization can be identified, based on contradictions within the System of Objectives. These can be solved either by compromise or by self-optimization. In the latter case, the System of Objectives is developed, and the List of Requirements and the function hierarchy are extended. In preparing the self-optimization concept, the self-optimization processes will eventually be defined, whether there is the necessary absence of conflicts of the self-optimization process will be analyze, and the boundary conditions within which self-optimization has to be working in, will be defined as well. For the newly integrated cognitive functions, solution patterns for self-optimization are identified and integrated into the partial models of the Principle Solution. The Principle Solution on the system level is the result of this phase.

To describe the system in detail, it is modularized and a Principle Solution for each single subsystem is developed in the "Conceptual Design on the Subsystem Level". This procedure corresponds to the Conceptual Design on the system level, starting out with planning and clarification of the task. This phase results in the Principle Solutions on the subsystem level. This process can be conducted recursively, because the subsystem itself can be a system with subsystems, and so forth.

The Principle Solutions of the subsystems are integrated into one Principle Solution during the phase "Concept Integration" to represent the whole system. Afterwards the thus-specified system is analyzed regarding its dynamical behavior and dependability. In this analysis phase, contradictions between the Principle Solution on the subsystem level are identified. Again, it must be determined whether these contradictions can be solved by self-optimization. At the end of this step, a technical-economical evaluation of the solution is made. The result of this phase is the Principle Solution of the whole system, which serves as a starting point for the subsequent Domain-specific Design and Development. This is carried out simultaneously in the specific domains (mechanical engineering, electrical/electronic engineering, control engineering and software engineering) [5].

In this early development phase, the consistency of the system is ensured by this domain-spanning approach. Development failures and therefore time-consuming and costly iteration loops can be avoided. To eliminate failures in this early stage, the dependability of the system is analyzed at the end of the concept integration. A number of methods can be used here to analyze the Principle Solution, such as the early Failure Mode Effect Analysis (FMEA) in combination with the Fault Tree Analysis (FTA) (cf. Sect. 3.1.1). To ensure the dependability of the self-optimization itself, the Multi-Level Dependability Concept can be applied (cf. Sect. 3.1.2). During the Design and Development, the Multi-Level Dependability Concept is realized in the domains where the failure has been identified (cf. Sect. 3.2.1). As long as failures or failure modes cannot be eliminated, the Principle Solution is tweaked and readjusted. The resulting Principle Solution forms the basis for the following Domain-specific Design and Development.

2.2 Domain-Specific Design and Development

As mentioned above, the Principle Solution is the starting point for the Domain-specific Design and Development. The Principle Solution contains the information that forms the basis for domain-specific development tasks. The transition of the Principle Solution to the domains involved is described briefly in *D.M.f.I.T.S*, [7], Sect. 5.1. In classical mechatronic development processes, the four domains mechanical, control, software, and electrical/electronic engineering are involved. For self-optimizing systems the optimization during operation needs to be additionally taken into account in the domains. The domains involved use their specific methods, tools and modeling languages to concretize the system. This phase is characterized by a high coordination effort; to ensure the consistency of the system, the results of the domains are continuously integrated. For this purpose, model transformation and synchronization techniques are used. The integrated system is tested as a virtual prototype to identify faults. This allows a short response time concerning design failures and therefore reduces time and cost-intensive iterations.

The reference process for the Design and Development of self-optimizing systems shows the ideal approach in which the particularities that need to be considered for the development of a self-optimizing system are pointed out. It is based on the development of the RailCab and its function modules. To reduce development time and therefore costs, the domains work in parallel where possible. Within the process, important synchronization points are depicted, where the domains exchange their results and get informations needed for further development. Even though the process gives the impression of being overly stringent, iterations can emerge in particular at these synchronization points, although they are possible at every stage of the process. The application of the presented approach for the Design and Development of a specific development task and company must be developed individually.

The approach presented here is recursive and conducted on different hierarchy levels of the system. The system itself consists of subsystems, whereby the subsystems are also systems themselves that consist of subsystems, and so forth. Fig. 2.4 presents a schematic representation of the process for one subsystem. (A detailed description is given in *D.M.f.I.T.S*, [7], Sect. 3.3.) It is intended to supplement the existing Design and Development process of the domains involved. In the following, we will give an overview of the tasks belonging to the various domains.

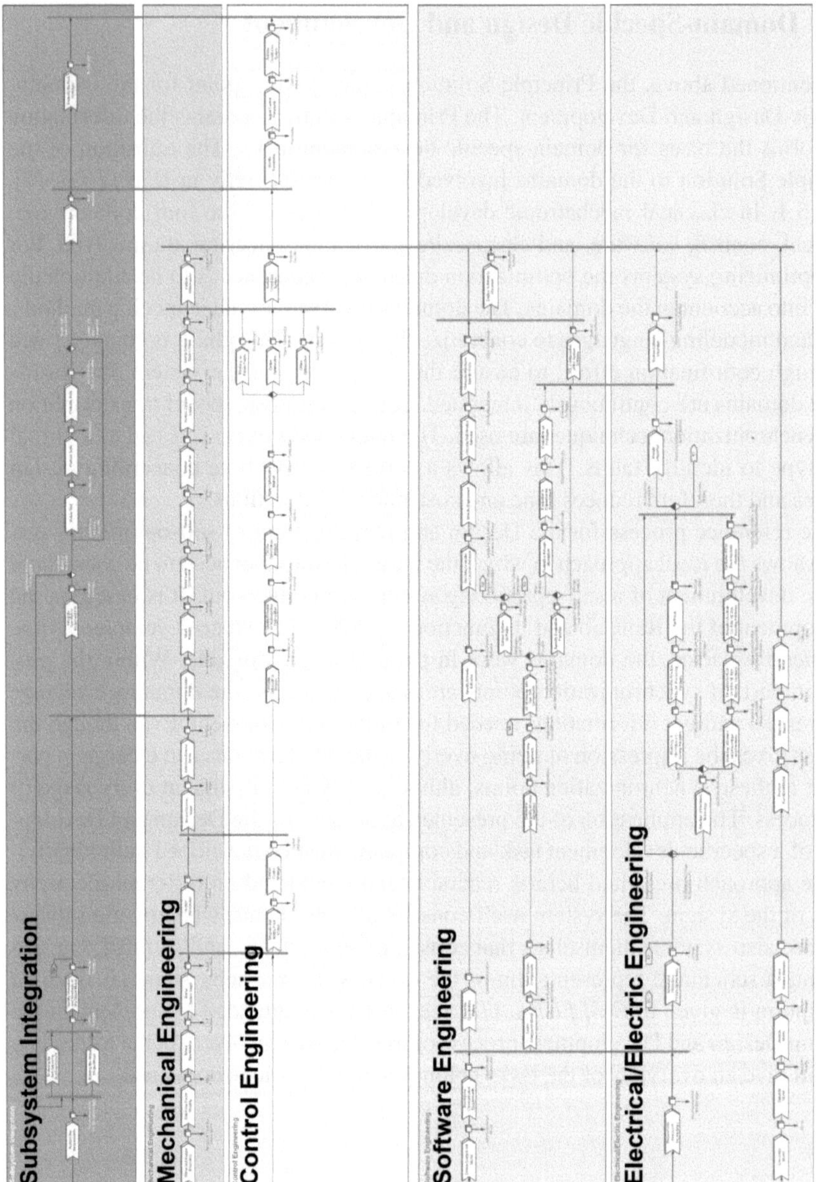

Fig. 2.4 Schematic representation of the Domain-specific Design and Development

Mechanical engineering:
The domain mechanical engineering has as its goal the Design and Development of the system shape of the self-optimizing system. Its starting point are the partial models Requirements, Active Structure and Shape of the Principle Solution. Based on these partial models, solution elements are selected and a first Shape model is built. This model serves as input for the domains control and electrical engineering, as well as for (sub)system integration. Using the Shape model, the domain control engineering is able to finalize the control strategy and the domain electrical engineering can develop a first model of the power electronics. Furthermore, a first virtual prototype can be modeled in the (sub)system integration. The result is a dynamic model that enables a simulation of the dynamic behavior. Based on these results, the Shape model can be further developed using a 3D-CAD software tool. The results of the work in this domain are the mechanical structure and derived manufacturing documents.

Control engineering:
The aim of control engineering is the design of the controller to guarantee the desired dynamical behavior of the self-optimizing system. Based on the Shape model from mechanical and electrical engineering, the control strategy can be finalized. The given dynamic model is the input for the optimization; resulting optimization strategies are integrated into the control strategy and the closed-loop system is analyzed. Thc analysis results are then passed along to the (sub)system integration and can be tested in the virtual prototype as hardware-in-the-loop-tests. Based on these tests, necessary parameters can be identified, control parameters adjusted, and the closed-loop can be analyzed again. These adjustments need to be done again after testing the real prototype; only then can the control strategy be finalized and implemented into the real system.

Software engineering:
The system software and discrete software are developed in the domain software engineering. The system software provides services for the discrete software; its development process is structured into three main phases. First, the required methods need to be identified and designed. Second, the developed methods need to be implemented; third they need to be evaluated.

In the first phase, the components of the system that will contain system software are selected. This information is derived from the Requirements of the Principle Solution. For each of these components, appropriate models or algorithms need to be determined if not already available. For example, self-healing may be implemented to improve the dependability of the system software (cf. Sect. 3.2.12). In the case of a high demand for processing power and memory, virtualization can reconcile the opposing requirements (cf. Sect. 3.2.15). Following this in the second phase, the new components need to be integrated into the system software and, if applicable, the system software is extended. If this extension is not possible, an iteration is necessary and the new approach needs to be adjusted or further models with their respective algorithms need to be developed. In the third phase, the system software components are assigned, including the specification of evaluation functions. The three phases are run through until all requirements are met.

During the Conceptual Design of the discrete software, a component structure is derived from the Requirements. The component structure of the discrete software is modeled by the Active Structure, while the communication between the components is modeled by Modal Sequence Diagrams.

During Design and Development, we use MechatronicUML (cf. *D.M.f.I.T.S*, [7], Chap. 5) to design the discrete software. In MechatronicUML, the component structure of the Active Structure is further refined and coordination patterns (cf. *D.M.f.I.T.S*, [7], Chap. 5) are derived from the Modal Sequence Diagrams; these coordination patterns specify communication protocols. Next, the discrete internal component behavior and the reconfiguration behavior are specified. We use model checking to guarantee that this behavior satisfies the requirements (cf. *D.M.f.I.T.S*, [7], Chap. 5); model checking also supports checking the integration of control engineering (cf. Sect. 3.2.10). The deployment of the software is based on the description of the hardware that is provided by electronic engineering. A hazard analysis is also performed at this stage to guarantee that the occurance of severe hazards remains below a specific probability limit. Hazards whose occurrence probabilities are higher may be reduced by self-healing operations, which are specified by means of reconfigurations. Then, it is analyzed whether the self-healing operations are able to achieve the intended probability reduction (cf. Sect. 3.2.8). Afterwards, the reconfiguration behavior is verified. The system computes reconfiguration plans at runtime that avoid unsafe configurations (cf. Sect. 3.2.9); this planning process is simulated at design time to validate its reliability. In addition, we apply online model checking (cf. Sect. 3.2.14). If all the requirements are fulfilled, the controller and the discrete behavior are simulated using the virtual prototype in the (sub)system integration (cf. *D.M.f.I.T.S*, [7], Chap. 5). If the simulation displays correct functionality, the hardware-specific code is generated and deployed on the hardware.

Electrical/electronic engineering:

Self-optimizing systems demand high flexibility in order to be able to adapt their behavior during operation. Dynamically reconfigurable hardware can ensure this flexibility. Furthermore, such systems demand high power transmission. Therefore, in the domain electrical/electronic engineering, both microelectronic devices for information processing and power electronics need to be developed. Input for the development of the microelectronics is given by the Requirements of the Principle Solution. After analyzing these requirements, a set of appropriate information processing hardware components is collected. In case the requirements cannot be fulfilled by existing hardware, additional hardware, for example embedded processor architecture or FPGA (Field Programmable Gate Array) technology is chosen. At this point the developer needs to decide whether the hardware to be developed consists only of static components, or of both static and dynamic ones. In the first case, the standard design flow from the Y-diagram for electronic engineering is used, whereas if dynamic reconfiguration of the hardware is considered necessary, the procedure needs to be extended with respect to the characteristics of these architectures. The concrete process is part of the design environment INDRA (Integrated Design Flow for Reconfigurable Architectures, cf. *D.M.f.I.T.S*, [7], Chap. 5. First, the dynamic reconfigurable hardware is modeled; while doing this, the requirements that are derived from system software are taken into account. The model is simulated and verified and afterwards synthesized until all of the components have been integrated. Finally, the system software can be ported to the new architecture. To increase the dependability of the system, self-healing via dynamic reconfiguration can be implemented (cf. Sect. 3.2.13). For the development of the power electronics, the partial model Requirements and Active Structure serve as input. Based on these, the type of electrical drive is selected and its capacity is analyzed. In the next step, the appropriate engine is chosen. This information is passed on to mechanical engineering and is integrated into the system model, after which the thermal resilience of the engine is tested and the data sheet of the electrical engine is derived. Afterwards, the power electronics can be defined. The results of these tasks are integrated into the overall system.

(Sub)System Integration:
To ensure the consistency of the domain-specific models, the results of the domains are integrated continuously into an interdisciplinary system model that is based on the Principle Solution; additionally the functionality of the system model is secured by a virtual prototype. To do so, a self-optimizing test bench and the environment are modeled. The basis for the virtual prototype are given by the Requirements of the Principle Solution. The virtual prototype is first modeled based on the basic Shape model developed by the domain mechanical engineering. This prototype is improved and expanded continuously over the course of development process by integrating the results from other domains. To make the simulation of the virtual prototype possible, the interactions and interfaces of the model need to be identified. For the system test, the test cases and a metric are first defined. Based on these cases, the virtual prototype is tested (cf. Sect. 3.2.3). The quality of the test results can be evaluated using the pre-defined metric. If the quality of the tests is not sufficient new tests are defined, performed and evaluated until the quality is sufficient. Then the test results are returned to the domains. When all subsystems are fully implemented into the virtual prototype and the test results are failure-free, the real prototype can be built and evaluated.

Within the development process, the dependability of the self-optimizing system needs to be continuously ensured. This can be done with dependability-specific methods such as those presented in Chap. 3. The Methodology for the Selection of Dependability Methods for the Development of Self-Optimizing Systems presented in Sect. 3.3 supports the developer in choosing the appropriate dependability method and in integrating it into the development process.

References

1. Albers, A.: Five Hypotheses about Engineering Processes and their Consequences. In: Proceedings of the 8th International Symposium on Tools and Methods of Competitive, Ancona, IT (2010)
2. Bender, K.: Embedded Systems – Qualitätsorientierte Entwicklung. Springer, Berlin (2005), doi:10.1007/b138984
3. Dorociak, R., Gaukstern, T., Gausemeier, J., Iwanek, P., Vaßholz, M.: A Methodology for the Improvement of Dependability of Self-Optimizing Systems. Production Engineering – Research and Development 7(1), 53–67 (2013), doi:10.1007/s11740-012-0425-3
4. Ehrlenspiel, K.: Integrierte Produktentwicklung, 3rd edn. Carl Hanser Verlag, Munich (2007)
5. Gausemeier, J., Frank, U., Donoth, J., Kahl, S.: Specification Technique for the Description of Self-Optimizing Mechatronic Systems. Research in Engineering Design 20(4), 201–223 (2009), doi:10.1007/s00163-008-0058-x
6. Gausemeier, J., Rammig, F.J., Schäfer, W. (eds.): Selbstoptimierende Systeme des Maschinenbaus. In: HNI-Verlagsschriftenreihe, vol. 234. Heinz Nixdorf Institute, University of Paderborn, Paderborn (2009)
7. Gausemeier, J., Rammig, F.J., Schäfer, W. (eds.): Design Methodology for Intelligent Technical Systems. Lecture Notes in Mechanical Engineering. Springer, Heidelberg (2014), doi:10.1007/978-3-642-45435-6_2
8. Isermann, R.: Mechatronische Systeme – Grundlagen. Springer, Berlin (2008), doi:10.1007/978-3-540-32512-3
9. Kahl, S., Gausemeier, J., Dumitrescu, R.: Interactive Visualization of Development Processes. In: Proceedings of the 1st International Conference on Modelling and Management of Engineering Processes (2010)
10. Pook, S., Gausemeier, J., Dorociak, R.: Securing the Reliability of Tomorrow's Systems with Self-Optimization. In: Proceedings of the Reliability and Maintainability Symposium, Reno, NV, US (2012)
11. Rieke, J., Dorociak, R., Sudmann, O., Gausemeier, J., Schäfer, W.: Management of Cross-Domain Model Consistency for Behavioral Models of Mechatronic Systems. In: Proceedings of the 12th International Design Conference, Dubrovnik (2012)
12. Verein Deutscher Ingenieure (VDI): VDI 2206 – Entwicklungsmethodik für mechatronische Systeme. Technical Guideline (2004)

Chapter 3
Methods of Improving the Dependability of Self-optimizing Systems

Abstract. Various methods have been developed in the Collaborative Research Center 614 which can be used to improve the dependability of self-optimizing systems. These methods are presented in this chapter. They are sorted into two categories with regard to the development process of self-optimizing systems. On one hand, there are methods which can be applied during the Conceptual Design Phase. On the other hand, there are methods that are applicable during Design and Development.

There are domain-spanning methods as well as methods that have been specifically developed for particular domains, e.g., software engineering or control engineering. The methods address different attributes of dependability, such as reliability, availability or safety.

Each section is prefaced with a short overview of the classification of the described method regarding the corresponding domain(s), as well as its dependability attributes, to provide the reader with a brief outline of the methods' areas of application. Information about independently applicable methods or existing relationships and interactions with other methods or third-party literature is also provided.

The development process for self-optimizing mechatronic systems which was introduced in Chap. 2 consists of two main phases: Conceptual Design and Design and Development. The main result of the Conceptual Design is the Principle Solution, which includes all information required for the concrete development during the second phase.

3.1 Conceptual Design Phase

Even as early as during the specification of the Principle Solution, the dependability of self-optimizing mechatronic systems can be evaluated and improved by employing appropriate methods. The result is an improved Principle Solution, as such methods take the whole system into account before splitting it up into domain-specific tasks for the following development phase.

J. Gausemeier et al. (eds.), *Dependability of Self-optimizing Mechatronic Systems*, Lecture Notes in Mechanical Engineering,
DOI: 10.1007/978-3-642-53742-4_3, © Springer-Verlag Berlin Heidelberg 2014

3.1.1 Early Probabilistic Reliability Analysis of an Advanced Mechatronic System Based on Its Principle Solution

Rafal Dorociak and Juergen Gausemeier

The Early Probabilistic Reliability Analysis of an Advanced Mechatronic System based on its Principle Solution is a method of improving the attributes reliability, safety and availability in the early development phase of Conceptual Design. The method can be used to ensure the dependability of the Principle Solution. Thus, it is necessary to use the specification technique CONSENS for the domain-spanning description of the system. Furthermore, the method for the early probabilistic analysis of the reliability of a self-optimizing mechatronic system uses two complementary reliability assurance methods, FMEA and FTA, in interplay. The method can be seen as a further development of FMEA and FTA to use both methods in the early phase of the development and in combination.

In the following, we will introduce the method for the early probabilistic analysis of the reliability of a self-optimizing mechatronic system based on its Principle Solution. This method allows for first statements regarding to the reliability of the system in the early engineering phase of Conceptual Design. In particular, the weak points of the system with respect to reliability are found. For those weak points, measures to detect and counter them are derived and implemented directly in the Principle Solution of the system. Altogether, the system under consideration is made more reliable at an early development stage.

3.1.1.1 Prerequisites and Input

The main input of our method is the domain-spanning specification of the Principle Solution. This Principle Solution is determined by means of the specification technique CONSENS for the domain-spanning [54] (cf. Sect. 2.1). As explained in Sect. 2.1, the description of the Principle Solution is divided into 8 partial models: Environment, Application Scenarios, Requirements, Functions, Active Structure, Shape, Behavior and System of Objectives. The focus of our method lies on the analysis of the partial models Environment, Application Scenarios, Requirements, Functions, Active Structure and Behavior.

3.1.1.2 Description

Following the recommendation of the CENELEC EN 50129 standard [45], the method for the early probabilistic analysis of the reliability of a self-optimizing mechatronic system uses two complementary reliability assurance methods FMEA (Failure Mode and Effects Analysis) [21,73] and Fault Tree Analysis (FTA) [21,74] in cooperation with each other. Some concepts known from the FHA (Functional Hazard Analysis) [150] method have been adapted as well, in particular, the use of a failure taxonomy for the identification of possible failures. By using these complementary methods, the completeness of the list of possible failure modes, failure causes and failure effects, as well as of the specification of failure propagation, is increased; both failure specifications are held mutually consistent.

Figure 3.1 shows the procedure model of our method iterations are not shown.

Phase 1 – specification of the Principle Solution:
The starting point of this phase are moderated workshops, where the experts from the involved disciplines work together in order to specify the system using the specification technique CONSENS, as well as to analyze and improve it with regard to reliability. In particular, the partial models Functions, Active Structure, and Behavior are described.

Phase 2 – early FMEA based on the Principle Solution:
The system structure and the corresponding functions are automatically derived from the description of the partial models Functions, Active Structures, and Behavior; they are recorded in the FMEA table. Failure modes, failure causes and failure effects are then identified. Checklists and failure taxonomies (e.g. as shown in Fig. 3.2) assist the failure identification process [47], [147]. In addition, combinations are identified of failure models which can conceivably

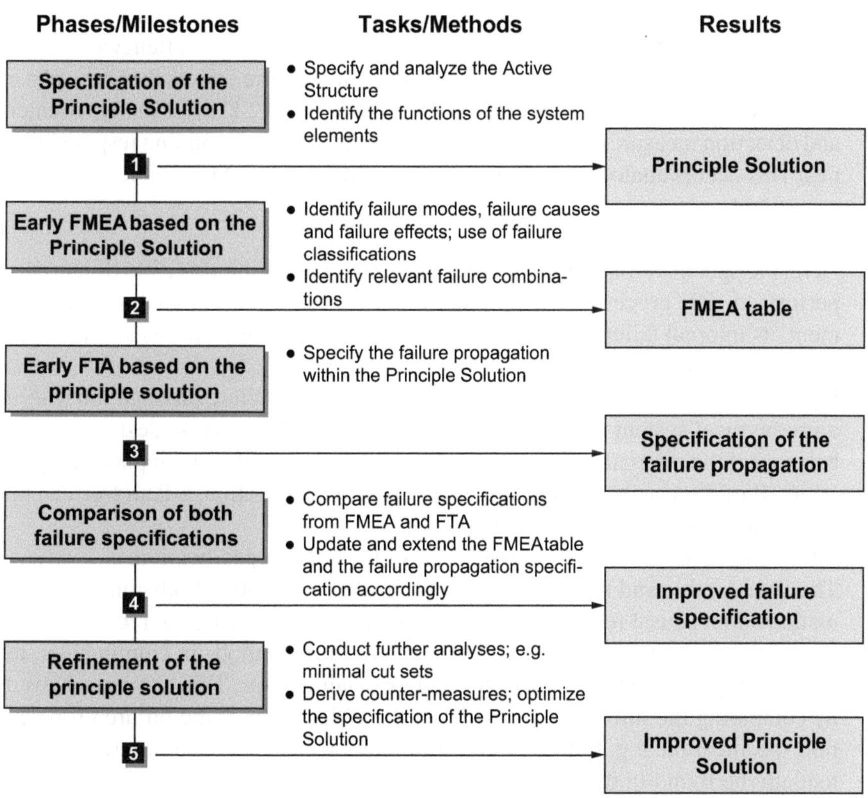

Fig. 3.1 The procedure model of the method for the early probabilistic analysis of the reliability of a self-optimizing mechatronic system

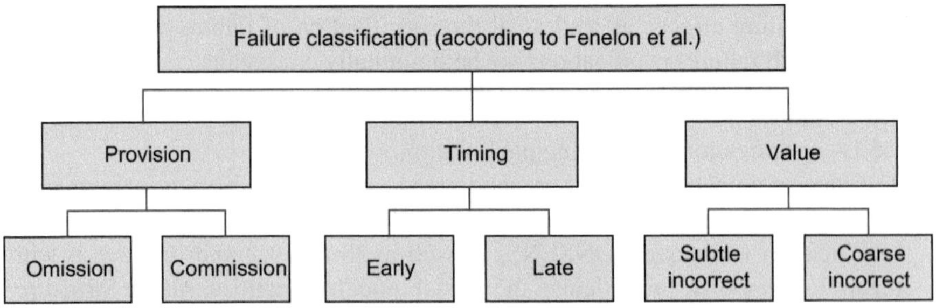

Fig. 3.2 Failure classification (according to Fenelon et al.) [47]

occur together and have a negative impact on the system (pairs of failures, triplets of failures, etc.). Failure modes and relevant failure mode combinations are recorded in the FMEA table. For each failure mode (and failure mode combination), the possible failure causes and failure effects are analyzed. Again, check lists can be used to accomplish this step, as they describe system elements known to be possible sources of problems with regard to reliability [43]. A number of failure effects can be found by analyzing the Principle Solution of the system; this, especially regarding the partial models Active Structure and Behavior. A risk assessment of the failure modes, failure causes and failure effects then take place using the risk priority number (cf. the norm IEC 60812 [73]). Finally, counter and detection measures are defined in addition as the corresponding responsibilities. This occurs analogously to the classical FMEA. The FMEA table is updated accordingly.

Phase 3 – early FTA based on the Principle Solution:

Here, the specification of the failure propagation within the Principle Solution is performed. The process is very similar to traditional FTA. For each system element, its internal failures as well as incoming and outgoing failures are specified and related to each other.

Figure 3.3 shows an example of the specification of the failure propagation within a prototypical system element SE. The output O_1 exhibits an undesired system behavior if the internal failure "failure1" occurs if one of the two inputs (I_1, I_2) is faulty. Based on such a description of the failure propagation, a fault tree can be generated (semi)-automatically (Fig. 3.3).

Phase 4 – improvement of the completeness of the failure specification:

The FMEA table and the specification of failure propagation both contain information about causal relationships between failures. Following the recommendation of the CENELEC EN 50129 [45], we use both methods in combination, to ensure a higher completeness of the failure specification. This can be achieved by comparing the information content of the FMEA and of the failure propagation specification: e.g. failures and causal relationships between failures can potentially be found in the failure propagation specification, which were not been found during the FMEA and are thus not documented in the FMEA table; the FMEA table is in that case updated accordingly. This also applies in the other

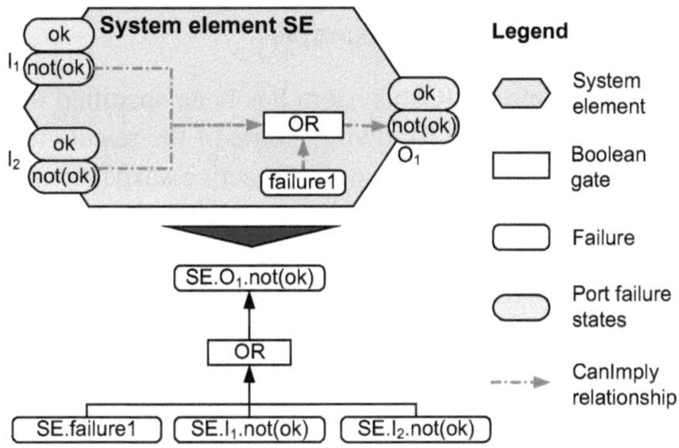

Fig. 3.3 Specification of the failure propagation and the corresponding fault tree

comparison direction: For example, if a causal relationship between two failures (e.g. between a failure mode and a failure effect) has been recorded in the FMEA table, there should be a corresponding causal relationship given in the failure propagation specification. If this is not the case, the causal relationship is incorporated into the failure propagation specification. During the process, additional failures can be found as well, which have not been specified at that point. The completeness of the identified failure modes, failure effects and failure causes, as well as of the failure propagation specification, is improved. Examples are provided at the end of this section, when our applied example is explained.

Phase 5 – Improving the Principle Solution:

Both failure specifications are analyzed. For instance, the classical analyses known from the FTA field, such as minimal cut sets, are used [21]. In particular, the importance analysis is performed. For this purpose, the Bayesian network-driven approach is used [38]; it enables the computation of the Fussell-Vesely importance measure. In such a manner, the most critical system elements are identified. Counter and detection measures are defined based on the analysis results. If possible, they are incorporated directly into the Principle Solution (e.g. redundancy, condition monitoring [90], etc.). Otherwise, they are recorded for further discipline-specific Design and Development (e.g. test and simulation measures, etc.).

3.1.1.3 Results

The result of this method is a revised Principle Solution of the system which is improved with regard to reliability. As a consequence, the system under consideration is made more reliable at an early development stage and a great number of time-intensive and costly iteration loops during the further development phases can be avoided. The failure specifications and analyses results from the Conceptual Design are used in the further development phase of domain-specific Design and Development. During this phase, with the increasing concretization of the system, reliability analyses such as FTA and FMEA are performed again.

3.1.1.4 Application Example

The complete RailCab system has been specified using the specification technique CONSENS. In the following, some of the results for the active suspension module of the RailCab are shown. Each active suspension module consists of three servo cylinders, which dampen vibrations and tilt the vehicle body in curves. Each servo cylinder consists of a hydraulic cylinder, a 4/4-way valve, a servo cylinder regulation and a hydraulic valve regulation [126]. The method for the early probabilistic analysis of the reliability of a self-optimizing mechatronic system has been applied to the active suspension module. As a first step, the Principle Solution of the active suspension module was modeled using the specification technique CONSENS. Figure 3.4 shows an excerpt of the partial model Active Structure of the servo cylinder that is used in the active suspension module.

Based on the specification of the Principle Solution, an early FMEA is performed. An excerpt of the resulting FMEA table for the servo cylinder is shown in Fig. 3.5.

Using the failure taxonomy by [47], the failure mode *hydraulic valve regulation provides no switch position for the 4/4-way valve* can be found. This failure mode occurs, for instance, if the energy supply of the system element hydraulic valve regulation is interrupted. According to the FMEA, the risk priority number for this case is 252. In order to eliminate or at least mitigate the failure mode, the energy supply of the hydraulic valve regulation should be monitored. One possible solution is the incorporation of an additional monitoring system element into the Principle Solution. In this case, additional measures, such as a redundant energy supply, have to be implemented.

Using the method for the early FTA, the specification of the Principle Solution is supplemented by the specification of failure propagation (Fig. 3.6). For each

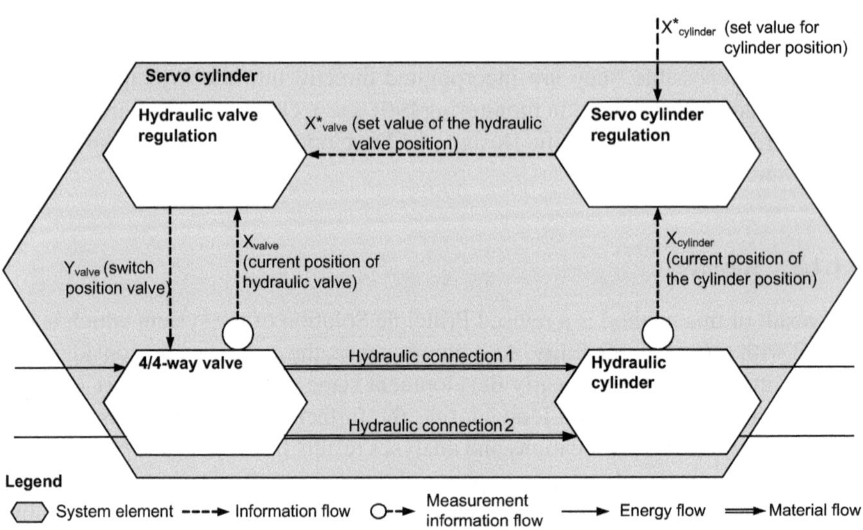

Fig. 3.4 Active Structure of the active suspension module (excerpt)

Failure Mode and Effects Analysis (FMEA)
Module: Servo cylinder

System element	Function	Failure mode	Failure effect	S	Failure cause	D	O	RPN	Counter- or detection-measure	...
Hydraulic valve regulation	Regulate position of the valve	Hydraulic valve regulation provides no switch position for the 4/4-way valve	Valve no longer changes pressure on output	6	Servo cylinder regulation does not set value for valve slider position	2	7	84	Monitor outgoing communication towards 4/4-way valve	...
					Hydraulic valve regulation broken	9	3	162	Generate warning message if needed	
					Energy supply of hydraulic valve regulation interrupted	7	6	252	Monitor energy supply	
4/4-way valve	Close valve	Valve closes in an unwanted manner	Hydraulic valve is stuck in current position	8	Energy supply interrupted	7	4	224	Monitor energy supply	...
					Magnetic coil damaged	8	3	192	Generate warning message if needed	
Servo cylinder regulation	Determine current position of cylinder piston	An incorrect value for the position has been determined	Control deviation $e = X^*_{cylinder} - X_{cylinder}$ has an incorrect value; the desired action is not generated	9	Sensor damaged	8	3	216	Monitor current value of cylinder lifting way $X_{cylinder}$	...
					Cable broken	9	2	162		
					Sensor wired in wrong way	3	2	54	Redundant design of measuring system	
					Wrong calibration	2	5	90		
...	...	...	...	...						

S: Severity of the failure effect
D: Detection probability of the failure cause
O: Occurence probability of the failure cause
RPN: Risk priority number (RPN=S·D·O)

Fig. 3.5 FMEA table of the servo cylinder (excerpt)

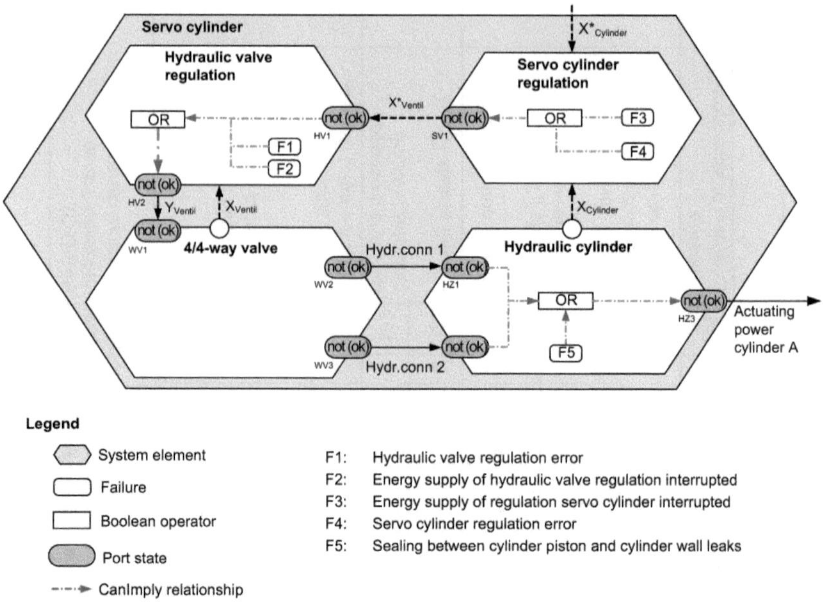

Fig. 3.6 Specification of the failure propagation of the servo cylinder (excerpt)

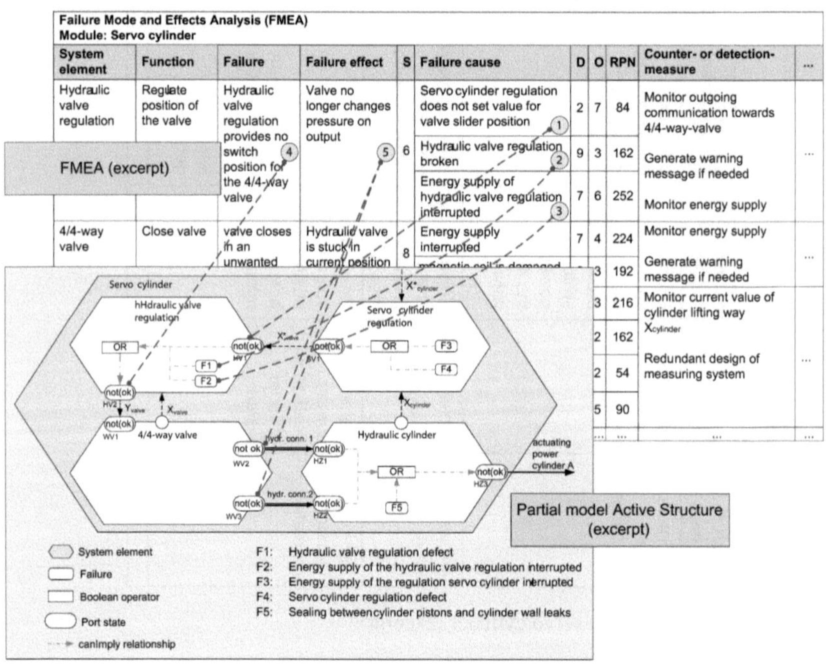

Fig. 3.7 Interrelation between the FMEA table and the specification of the failure propagation

system element, the relationship between incoming, local and outgoing failures is described. Figure 3.7 depicts the interrelation between both failure representations. The failure cause *servo cylinder regulation does not provide desired position of valve switch* from the FMEA table (Fig. 3.7, (1)) corresponds to the port state not(ok) of the input HV1 of the system element hydraulic valve regulation. The failure causes *hydraulic valve regulation is broken* (2) and *energy supply of the hydraulic valve regulation is interrupted* (3) correspond to the internal failures F1 and F2 of the hydraulic valve regulation. The aforementioned failure causes (2) and (3) may lead to the failure *hydraulic valve regulation provides no switch position for the 4/4-way valve* (4); This failure is recorded in the FMEA table, as well as in the specification of the failure propagation (port state not(ok) of the output HV2 of the hydraulic valve regulation).

According to the FMEA table, there is a causal failure relationship between the failure *hydraulic valve regulation provides no switch position for the 4/4-way valve* (4) and the failure effect *valve no longer changes the pressure on the output* (5). Although both failures were specified in the failure propagation model (input WV1 of the 4/4-way valve as well as inputs HZ1 and HZ2 of the hydraulic cylinder, respectively), the causal relationship between them had not been modeled; in a consequence, a more thorough analysis of this causal relationship was performed. In the course of the analysis, the respective failure propagation path was modeled, as well as an additional failure F6 (*valve position can no longer be changed mechanically; the valve slider stays in its current position*) (Fig. 3.8).

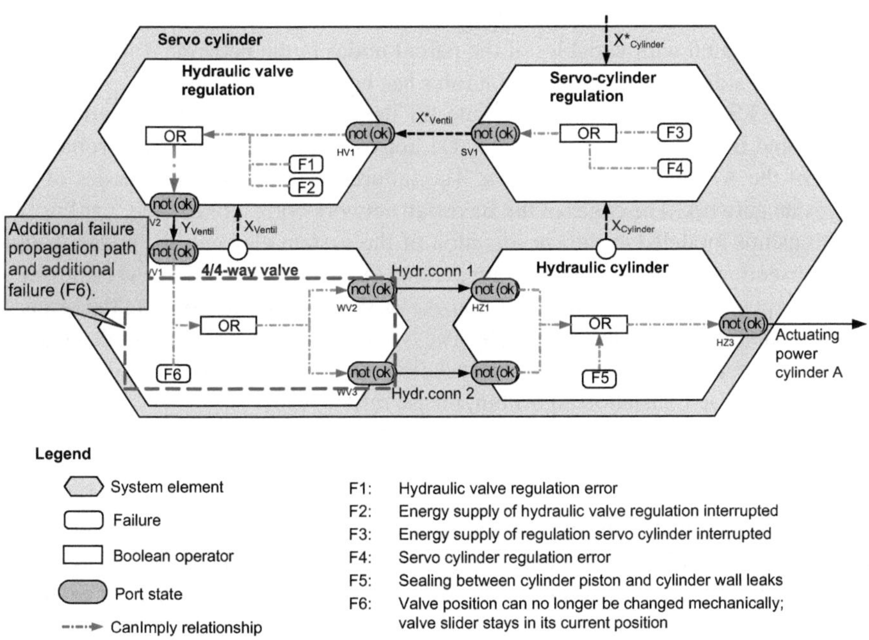

Fig. 3.8 The extended failure propagation specification of the servo cylinder

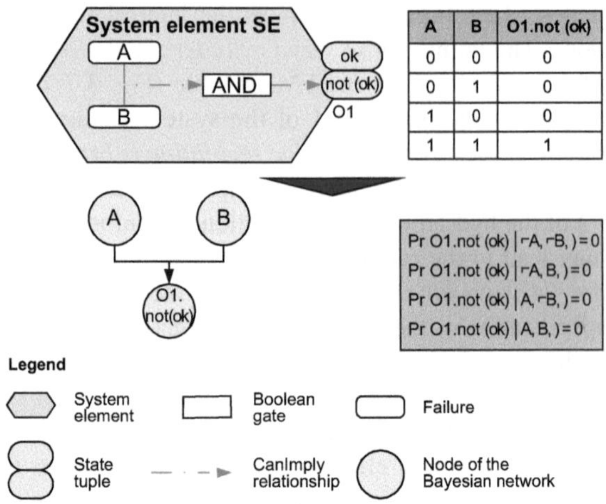

Fig. 3.9 Excerpt from the translation dictionary [38]

The specification of the failure propagation is translated into a Bayesian network. The translation algorithm proceeds as follows: for each system element, its internal failures and port states of its inputs and outputs are translated into nodes of the Bayesian network. The relationships between them are represented as edges in the Bayesian network. The Conditional Probability Table (CPT) of the Bayesian network is then populated: for each value of variables associated with a node or a node state, its conditional probabilities are described with respect to each combination of values associated with variables of the parent nodes in the network. To support the translation, a dictionary of translation rules has been developed [38].

Figure 3.9 shows the translation rule for the AND gate with two internal failures A and B and the outgoing failure "O1.not(ok)". The corresponding probability table of the AND gate is also shown. The failures are translated into nodes of the Bayesian network. The edges of the Bayesian network correspond to the "canImply" relationships modeled in the specification of the system element SE: nodes A and B representing the internal failures are parents of the outgoing failure "O1.not(ok)". An excerpt of the CPT for node "O1.not(ok)" is shown. It corresponds to the respective probability table of the AND gate. The translation rules for other Boolean gates are analogue. For other examples of translation rules and a more detailed description of the respective translation algorithm, please refer to [38].

The result is a comprehensive Bayesian network which describes the part of the system that is relevant for the examination of the chosen top event. Based on the Bayesian network representation, some further analyses are performed [89]. In particular, the Fussell-Vesely importance measure is computed, i.e. it is determined with what probability a particular system element (failure cause) had led to a particular failure (the so-called posterior probability). The top event that we examine is *valve no longer changes the pressure on the output* (corresponds to the port state WV2.not(ok)). Let us consider the state of the failure specification before the additional failure F6 and the corresponding propagation path were incorporated into the

Table 3.1 Failures, the failure rates and the Fussell-Vesely importance measure (before and after the failure specification had been extended) (Top-Event is WV2.not(ok))

Failure	Failure rate (per hour)	Fussel-Vesely importance (before)	Fussel-Vesely importance (after)
F1	5.11×10^{-7}	0.2897	0.2416
F2	4.02×10^{-7}	0.2279	0.1901
F3	3.28×10^{-7}	0.1859	0.1551
F4	5.23×10^{-7}	0.2965	0.2473
F6	3.51×10^{-7}	N/A	0.1660

specification. The failure rates of the failures are shown in Tab. 3.1. We will further assume that the output WV2 of system element hydraulic valve regulation is in state not(ok), as this is our top event. According to the specification of the failure propagation (Fig. 3.6) failures F1, F2, F3 and F4 all contribute to this. Table 3.1 reports Fussell-Vesely importance measures for each failure, i.e. the posterior probability of the contributing failures given the occurrence of the aforementioned failure. Failures F1 and F4 are especially important, with importance greater than 28 %.

Now let us consider the extended specification of the failure propagation (including failure F6). The failure rate of failure F6 and the respective importance measures are shown in Tab. 3.1. The failures F1, F2, and F4 are of highest importance with importance with an approximately 25 %.

By using our method, the completeness of the failure specification has been improved. In particular, failures and failure relationships were identified that could easily have been omitted otherwise. In particular, the failure F6 has been identified, the importance of which is quite high (approximately 17 %). Based on the failure specification, further analyses are conducted. Counter and detection measures are then derived and, if possible, implemented directly in the Principle Solution. Altogether, the system under consideration is made more reliable at an early development stage.

3.1.2 Early Design of the Multi-Level Dependability Concept

Rafal Dorociak, Jürgen Gausemeier, Tobias Meyer, Walter Sextro, and Christoph Sondermann-Woelke

The Multi-Level Dependability Concept (M-LD Concept) is an approach for improving the dependability, specifically the attributes reliability, safety and availability, of a self-optimizing system by using self-optimization. It is advantageous to develop the M-LD Concept within a single task once the Principle Solution has been fully specified and before the domain-specific phase begins.

To this end, this method has been developed, which allows a structured setup of the M-LD Concept based on the Principle Solution. It is a modification and an expansion of the procedure described in ISO 17359 [1, 140]. Since it is a multidisciplinary approach, information from several partial models (Active Structure, System of Objectives, Behavior, Functions, Environment and Application Scenarios) is required to best use the full potential of a self-optimizing system.

3.1.2.1 Prerequisites and Input

Since changes in late development phases usually come at great expense, it is advantageous to develop and insert the M-LD Concept into as early a development phase as possible. To do so, information from the Principle Solution is required; therefore, the system needs to be completely specified by a Principle Solution according to *D. M.f.I.T.S*, [55], Sect. 4.1.

Additionally, the system's information processing must be set up as an Operator Controller Module (see also Sect. 1.1.1) in order to be able to carry out the interactions between M-LD Concept and the system itself. The OCM already contains a preliminary Configuration Control, which can be improved using this method. In order to adapt the system behavior, multiobjective optimization must be feasible and it must be possible to implement additional objective functions, as will be discussed in Sect. 3.2.1.

To obtain information about the system itself, an FMEA (see also [73]) is required, which can be conducted according to the method *Early Probabilistic Analysis of an Advanced Mechatronic Systems based on its Principle Solution*, see Sect. 3.1.1. Since this method requires the Principle Solution as well, it does not pose further challenges.

3.1.2.2 Description

The basic approach is outlined in Figure 3.10. As shown, there are five phases with corresponding milestones which will be explained in detail. Phases and steps are not to be seen as a sequence; the procedure is characterized by a number of iterations, which are not depicted.

Step 1: System Analysis

The system analysis is conducted in three steps: analysis of the current system design, determination of the relevant system's objectives, and identification of possibilities of adapting the system behavior during operation. The required pieces of information and the relations between them have already been in the Principle Solution, so it is natural to use the partial models Active Structure, System of Objectives and Behavior for this analysis.

For the first step of the system analysis, the partial model Active Structure is used. It describes the system elements chosen to fulfill the required functions of the system. To obtain an overview of the system's capability to monitor its momentary state, a list of all sensors and the corresponding measurement points is generated by the program Mechatronic Modeller, which is used to model the Principle Solution.

In the second step of the system analysis, the objectives that are relevant with regard to the dependability of the system are identified. The objectives regarding

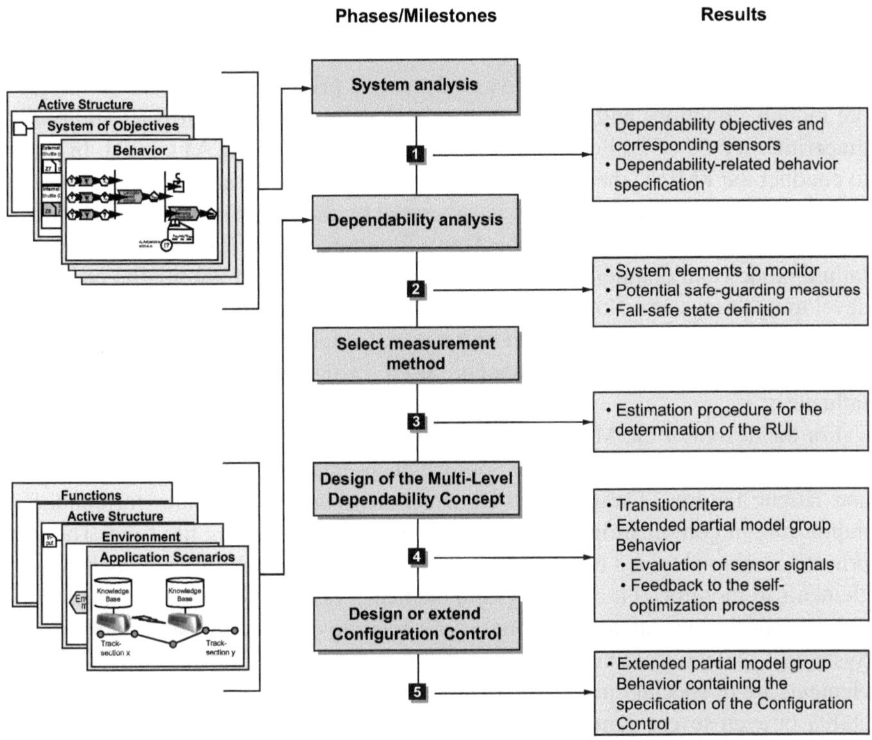

Fig. 3.10 Procedure for the design of the Multi-Level Dependability Concept; information contained in the partial models of the Principle Solution serves as input, as depicted by the partial models in the left column

dependability, e.g. "Maximize reliability", "Minimize down-time", "Minimize wear", which are influenced later on by the M-LD Concept, are extracted from the partial model System of Objectives. In most cases, these objectives are not as obvious as stated above; thus, the relevance of each objective concerning reliability needs to be evaluated. Afterwards, the previous list of sensors is examined in light of the dependability-oriented objectives to determine which sensors are to be used to determine the current system state and how they should be classified in the M-LD Concept.

The third step is to identify possibilities of influencing the system behavior. The relevant partial model group Behavior illustrates the different system states and reconfiguration options. These states and reconfiguration options will be part of the Configuration Control, which is designed to switch to the desired control strategy. The behavior specifications that support dependability-oriented actions will be used during Step 5.

Step 2: Dependability Analysis

The second step – the dependability analysis – is primarily conducted on the partial models Functions and Active Structure. Usually, the established reliability engineering method of Failure Mode and Effects Analysis (FMEA) is used. In order to conduct the FMEA, the system elements of the Active Structure are exported to a FMEA tool [39]. This procedure is supported by the program Mechatronic Modeller, as it is capable of exporting the Active Structure model. Based on this data, the failure modes and counter-measures are determined together among the engineers developing the system. Not only effects inherent in the system, but also effects from the surrounding environment could be a reason for failures. The partial models Application Scenarios and Environment are used for the process of identifying these influences.

For the design of the M-LD Concept, several results from the FMEA are important. Firstly, the failure modes point out which system elements are subject to wear and fatigue failures. These system elements are primary candidates for condition monitoring in combination with an estimation of the remaining life time. The risk priority number obtained from the FMEA is crucial for the decision of which system elements to monitor. If a critical system element's failure mode is not related to an objective of the system, an additional corresponding objective must be added to the System of Objectives. Furthermore, counter-measures also indicate which system elements are of special interest due to the fact that a failure would lead to low availability or even severe damage. The counter-measure list also shows failures which lead to a state in which operation of the system is still safe, thus forming the fail-safe state, which has to be defined for any self-optimizing system. Finally, failures of those system elements which have a negative influence on the dependability-oriented objectives are identified. For these, safeguarding against failures, e.g. redundancy, might be required [75]. If safeguards are used, the Principle Solution has to be updated accordingly.

Step 3: Select Measurement Method

In the system and dependability analysis, sensors for monitoring the dependability-related objectives as well as critical system elements are identified. Based on this information, the dependability-oriented objectives are related to quantifiable general measures, such as the remaining useful life (RUL) or the current failure probability of the system. These general quantities simplify the comparison between different system elements and subordinated system elements. The estimation of the RUL can be based on model-based approaches in which the actual load of the system elements is compared to the maximally tolerable load. In combination with damage accumulation hypotheses, e.g. Palmgren-Miner [103], the RUL can then be estimated. If desired, the failure probability of the system elements is also calculated using the corresponding distribution functions.

Step 4: Design of the Multi-Level Dependability Concept

The M-LD Concept is intended to influence system behavior; for this, an evaluation of the current system state and feedback to the system is needed. The system behavior can be influenced in two ways: by adapting the prioritization of the objectives of the system and/or by switching between different control strategies.

To determine whether the objectives need to be adapted, thresholds between four levels, based on the RUL or other general criteria, as selected in Step 3, need to be defined. For this, the safety requirements of the module have to be taken into account. As will be explained in Sect. 3.2.1, the dependability-oriented objectives, which support different attributes of the dependability such as reliability, availability, and safety (cf. [11]), are adapted if the second level is reached. Should the third level be reached, the objectives for safety must have absolute priority. In both cases, the system is influenced later on by an increase in priority of the dependability-related objective, which leads to more dependable operation. In order to increase the priority, a suitable fixed value for the priority or a strategy to increase it has to be implemented in the Cognitive Operator. Both evaluation of the sensor signals and feedback to the optimization process are integrated into the partial model group Behavior.

If a failure requires a switching action, this is set into motion by the Configuration Control; how exactly to implement this is explained in Step 5. The fourth level corresponds to the fail-safe state determined in the dependability analysis. If it is reached, emergency routines are engaged.

Step 5: Design or Expand Configuration Control

Certain failures (identified by the FMEA) could lead to a switch in control strategy, e.g. if a required sensor fails and redundancy controls require the system to switch to another sensor signal. This reconfiguration is conducted by the Configuration Control, which is embedded into the Reflective Operator. For the reconfiguration, different control strategies are designed. If switching actions are necessary, they have to be included in the Configuration Control. Since switching actions have to be initiated quickly, the required failure detection methods are implemented in the Configuration Control as well. The preliminary Configuration Control included in the partial model group Behavior is expanded to include dependability aspects.

3.1.2.3 Results

The M-LD Concept is fully specified and its components are embedded into the corresponding partial models of the Principle Solution. These components are both new system elements, i.e. hardware, as well as additional components in the information processing, which are mainly embedded into the Reflective Operator in the Operator Controller Module. The determination of the current system state, its classification according to the four levels and the initiation of the corresponding counter-measures are included in the partial model group Behavior. The Configuration Control is expanded or initially designed using this method and new dependability-related objectives are included in the multiobjective optimization.

3.1.2.4 Application Example

As a demonstrator for this method, the Active Guidance Module (see also Sect. 1.3) has been selected. It is a key element of a RailCab and, as such, needs to function dependably. To ensure this, the M-LD Concept has been implemented. The execution of the five main steps to design the M-LD Concept, as described in Sect. 3.1.2.2, is explained in the following.

Step 1: System Analysis

To analyze the current situation, the Active Guidance Module is equipped with several sensors. One incremental sensor at each wheel determines the longitudinal position of the RailCab. Since a drift of the incremental sensor's signal is unavoidable, the longitudinal position is regularly corrected by a proximity switch which passes over a reference plate. Furthermore, eddy-current sensors on each side of each wheel are used to measure the current lateral position as the deviation from the center line within the track, as well as the current clearance which could be used for optimization of the trajectory within the track limits. Two acceleration sensors and a yaw rate sensor are integrated into the construction in order to obtain further information about the RailCab movements. A displacement sensor is integrated into the hydraulic steering actuator.

The Active Guidance Module uses multiobjective optimization for the steering control strategy. The qualitative representation of the optimization objectives are given in the partial model System of Objectives; an extract is depicted in Fig. 3.11. The main goal is to steer within the track clearance while neither having flange contacts nor wasting energy on unnecessary steering actuator movements. At the same time, lateral acceleration has to be kept low to ensure passenger comfort and to be sure a certain safety margin is upheld [56].

The main dependability issue is the minimization of flange contacts in order to increase the reliability and thus the availability of the RailCab. Another objective is to minimize the wear of the hydraulic actuator. This is similar to the objectives "Maximize comfort" and "Minimize energy consumption", since all three lead to minimal actuator movements. For this reason, no additional objective is required.

The configuration control in the Active Guidance Module used for this example comprises several control strategies. The most advanced strategy uses the optimization described above and both a feedforward controller as well as a feedback controller. If no optimization is available, the trajectory generated for the feedforward controller is oriented towards the center line between the rails. If the eddy-current sensors fail, and with these the determination of the lateral position, the feedforward control can still be used to keep the vehicle on the center of the track. If all systems fail, the steering will become stuck, which leads to rapid wear on the flanges. In this case, the mechanical guidance wheels are activated and the vehicle is slowed.

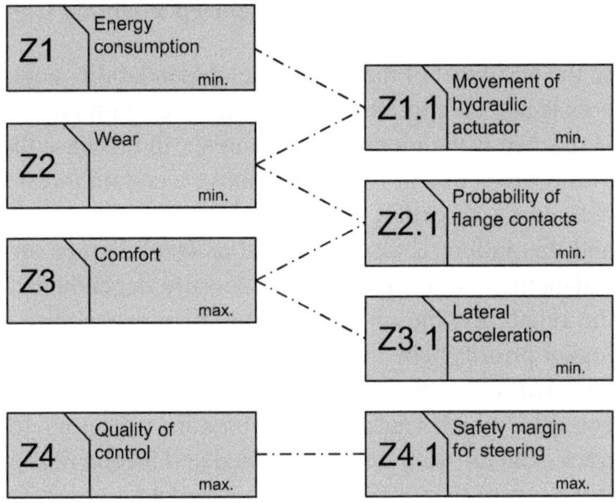

Fig. 3.11 System of Objectives of the Active Guidance Module (excerpt)

Step 2: Dependability Analysis

For the active guidance module, the system elements to be monitored are the wheels, since the rolling contact leads to wear. Besides the continuous wear due to the unavoidable motion between wheel and rail, flange contacts increase wear considerably. This is represented in the objective "Minimize probability of flange contacts". These contacts cannot be avoided completely, since steering along energy-efficient trajectories requires reducing the safety margin as much as possible, making flange contacts highly probable if unexpected disturbances are encountered.

By taking the Application Scenarios into account, it becomes obvious that severe accidents could occur while going over passive switches. Therefore, two different counter-measures are integrated into the partial model Behavior. The first counter-measure is the fail-safe state. In this state, the steering axle is fixed, if possible in center position; the velocity of the RailCab is reduced; and mechanical guidance wheels for going over passive switches are engaged. The second counter-measure is to safeguard the eddy-current sensors.

Step 3: Select Measurement Method

To determine the wear of the wheels, flange contacts are registered by the eddy-current sensors and the distance travelled is monitored via the incremental sensors. The maximum running length of the flanges in contact is compared to this value to obtain the RUL. The wear of the actuator is assumed to be proportional to its total distance travelled, which is calculated using the displacement sensor integrated into the actuator.

Step 4: Design of the Multi-Level Dependability Concept

In the first level of the multi-level dependability concept, the self-optimization process is able to choose from all objectives without any constraints. The second level is reached if the monitored parameter, in this case the rate of reduction of the RUL of the wheel due to wear, rises above a certain threshold which has been previously defined by an expert. If one of the eddy-current sensors fails, redundancy is lost, and this failure is also classified as level 2 since an error has occurred. However, self-optimization can be used to ensure dependable operation. In order to increase the reliability, the objective "Minimize probability of flange contacts" receives a higher priority. The third level is reached when the loss about the lateral position is detected. It is now of paramount importance to minimize the probability of flange contacts, which leads to a feedforward trajectory following the center line of the track. The fail-safe state "axle fixed and mechanical guidance activated" is activated if the loss of the longitudinal position data is ascertained. Purely closed loop control is not possible, since passive switches would then lead to a derailment.

The partial model group Behavior is extended to include the evaluation of the sensor signals, as described in Sect. 3.1.2.4, in addition to all required switching actions or adaptations of the objectives.

Step 5: Design or Expand Configuration Control

For the Active Guidance Module, switching is required if one of the redundant eddy-current sensors fails. If this is the case, the failed sensor's signal has to be neglected, requiring a switch to a different sensor evaluation algorithm and possibly to another control strategy. Both the detection of the failure as well as the switching process are embedded within the Configuration Control.

The final Configuration Control is based on a preliminary Configuration Control, which is already in place for general steering purposes. The additional actions expand the partial model group Behavior.

When all required components have been included in the Principle Solution, the design of the Multi-Level Dependability Concept is concluded.

3.2 Design and Development

In the phase Design and Development, particular emphasis is laid on software. Self-optimizing mechatronic systems contain discrete software as well as continuous software. Continuous software is used in control engineering, discrete software in software engineering. We use a component-based approach to develop the systems; the component structure is derived from the Active Structure of the Principle Solution and provides the basis for the methods presented below.

3.2.1 Increasing the Dependability of Self-optimizing Systems during Operation Using the Multi-Level Dependability Concept

Jan Henning Keßler, Tobias Meyer, Walter Sextro, Christoph Sondermann-Woelke, and Ansgar Trächtler

Self-optimizing systems offer the possibility of enhancing system dependability by adapting the system behavior to the current level of deterioration. An adaptation of the system behavior can be used to reduce the loads on individual system components, e.g. actuators, in order to make them less prone to failure. As this can be carried out during operation, a significant increase in reliability and usable lifetime or a limitation of the risk during operation can be achieved, thus improving the attributes reliability and availability or safety. However, implementing the necessary behavioral adaptation is challenging. The concept introduced in the previous section, the M-LD Concept, can be used to overcome these challenges and adjust system behavior during operation by applying self-optimization specific counter-measures, such as an adaptation of the objectives of the system, which are included in the partial model System of Objectives, and system reconfiguration.

3.2.1.1 Prerequisites and Input

The M-LD Concept has been developed to influence the behavior of a self-optimizing system during operation. To this end, it uses self-optimization as a tool to adapt objectives and to initiate switching actions, e.g. to switch between different controller configurations. The M-LD Concept is designed for the Operator Controller Module information processing architecture, see also Sect. 1.1.1, and has to be embedded into the Reflective Operator. As part of the Reflective Operator, the M-LD Concept can influence Self-Optimization on the Cognitive Operator level and also initiate switching actions by interacting directly with the Configuration Control.

For the framework of the M-LD Concept, model-based self-optimization is used. It is based on multiobjective optimization for calculating optimal system configurations, which consist of certain parameters that set the working point (see also Sect. 3.2.11). In order to use multiobjective optimization, a model of the dynamical behavior of the system is required, which can be gleaned from the partial models Environment, Application Scenarios and Active Structure. To adapt the system behavior, debendability-related objective functions, which are included in the partial model System of Objectives, need to be incoporated into the multiobjective optimization. These have to be defined so that their prioritization results in more dependable system behavior.

3.2.1.2 Description

The M-LD Concept can be developed and included in a self-optimizing system to increase the dependability of the system. It contains four hierarchically ordered levels for the characterization of the deterioration of the system (see Fig. 3.12) and is part of the Reflective Operator.

For each level, certain counter-measures affecting the system behavior have to be defined. To adapt the system's behavior using self-optimization, the priority rankings of the system's objectives are modified. The objectives in turn influence the behavior and thus the dependability of the system. The levels of the dependability concept and the resulting change of the system behavior are as follows:

Level I:
: The system operates dependably. Dependability is one objective among others; no counter-measures are required.

Level II:
: A minor error has occurred. Self-optimization is used to ensure dependable operation. The priority of the dependability objective affected by the error is increased, altering the system behavior. If the priority of a dependability-related objective is increased, the system behavior becomes more dependable, but at the same time other objectives must be subordinated.

Level III:
: A severe error has occurred, but the system can still be controlled. First emergency mechanisms are triggered to achieve a safer state. Of all the objectives, those objectives that lead to safe behavior become the primary objectives in order to avoid the failure of the whole system and the consequences involved. If there are separate objectives, the other attributes of dependability (e.g. reliability, availability) may occur as secondary objectives after those that are important to ensure safe operation. It is also possible for the concept to execute switching actions, e.g. to deactive failed system components.

Level IV:
: Control over the system is lost. Emergency routines are executed to reach a predefined fail-safe state.

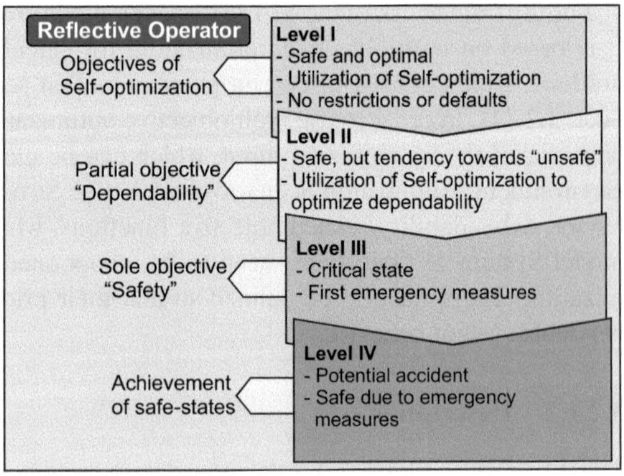

Fig. 3.12 The four levels of the Multi-Level Dependability Concept

As a component of the Reflective Operator, the concept is both able to get sensor information from the Controller layer of the Operator Controller Module and to communicate directly with the Cognitive Operator to influence the optimization process of the system. Situated in the Reflective Operator, the concept is also able to initiate switching operations between different control strategies via the Configuration Control.

3.2.1.3 Results

The M-LD Concept is set up to increase the dependability of a self-optimizing system during operation. Additional components, which allow a classification of the current system deterioration into four levels, are added to the Reflective Operator. According to these levels, counter-measures are initiated. These are either an adaptation of the system behavior via self-optimization, i.e. an adaptation of the priorities of the objectives to suit the current situation, or switching to a different controller strategy.

3.2.1.4 Application Example

To show the interaction of the M-LD Concept with the other components of the system, the Active Suspension Module (see also Sect. 1.3) is used as an example. The purpose of this module is to generate additional damping forces between the chassis and the body to actively control body motion. It consists of several major parts: A body framework, which represents the vehicle's body mass, and two actuator modules, each with one glass-fiber-reinforced plastic-spring (GRP-spring) mounted symmetrically beneath. The GRP-springs are connected via sophisticated nonlinear guiding kinematics to the coach body and, at the lower end, rest on the excitation unit representing the vehicle's axle, where three hydraulic cylinders actively displace each spring base. With its six actuators and three degrees of freedom in vertical, horizontal and rotational direction, the actuated system is over-determined. The actuator redundancy is an important feature in increasing the system's dependability. However, due to the nonlinear kinematics, control reconfiguration is required in case of faults.

The main failure mode is a fault in one or more of the six hydraulic actuator modules, which each consist of one servo valve and the corresponding linear actuator, that inhibits oil flow in the affected system. As a result, the corresponding actuator becomes stuck. To be able to continue controlling the body motions in case of such a fault, redundancy had to be implemented.

The four levels of the M-LD Concept are described in detail as part of the design process. Level I, being the nominal case, does not require any further action. Level II is reached when the estimated remaining useful life of the actuators falls below a pre-defined threshold. To be able to use self-optimization to increase the dependability, an objective corresponding to the actuators' state of deterioration needs to be defined. If the remaining useful life of one of the actuators becomes critically low, Level III is reached. To avoid the system becoming uncontrollable, the critical actuator is deliberately deactivated at this stage, whereby its function is compensated by the remaining actuators. However, the potential of the parallel redundancy given by the six actuators cannot be used with a conventional controller design, since a conventional controller would not be able to handle the change in system behavior and would thus become unstable. In order to maintain desired system behavior in case of actuator failures, control reconfiguration is used. Level IV corresponds to the fail-safe-state and is reached if more actuators fail than can be compensated for. In this case, the active suspension system is shut down and, to prevent dangerously high excitations, the vehicle's drive system is restricted to a low speed.

During normal usage, i.e. if the current system state is classified as Lvel I of the M-LD Concept, the active suspension system pursues two main objectives: on one hand, the energy consumption of the Active Suspension should be minimized, and on the other hand, passenger comfort should be maximized. This is achieved by minimizing the *dis*comfort, which corresponds to vertical and lateral accelerations of the body [71]. These two objectives conflict, as a reduction in body accelerations results in a higher energy consumption. The corresponding objective functions are:

$$f_{1,E} = \frac{1}{T} \cdot \int_0^T \sum_{j=1}^{6} P_{hyd,j}(\tau)\,\mathrm{d}\tau, \tag{3.1}$$

$$f_{2,comf} = \frac{1}{T} \cdot \int_0^T \sum_{i=1}^{2} |W_i(a_i(\tau))|\,\mathrm{d}\tau. \tag{3.2}$$

Equation 3.1 describes the average hydraulic power of the six actuators $P_{hyd,j}$; Eqn. 3.2 describes the comfort with reference to the weighted body accelerations a_i. The accelerations are weighted according to [148] to represent the subjective perception of a passenger.

If individual actuators show signs of wear that cannot be tolerated at the current operating time, Level II is reached and the self-optimization procedure chooses a new Pareto optimal point, which focuses on a higher dependability, as explained in Sect. 3.2.11. This requires a third objective taking dependability into account, called "minimize *un*dependability". For the given system, this objective needs to relate the current controller configuration to the resulting deterioration of the hydraulic valve, taking the characteristics of hydraulic valve deterioration into account. In order to reduce the rate at which the actuators deteriorate, this objective is prioritized over the others. However, if the rate cannot be reduced sufficiently using the results of the multiobjective optimization, the least reliable hydraulic cylinders have to be shut off. This leads to a change of the structure and the dynamic behavior of the system and usually to an unstable behavior of the closed-loop system if using the standard control structure. Control reconfiguration is used to keep the system operational by taking advantage of the redundancy and using the remaining five cylinders.

In order to adapt the system behavior, the objective "minimize *un*dependability" has to be defined and integrated into the optimization problem. It has to take the mean amount of motion of all actuators into consideration as well as considering the possibility of an uneven load on the actuators which could lead to premature failure of individual actuators. A prioritization of the objective corresponding to this function has to lead to less motion of the actuators, which increases their dependability. Unfortunately, at the same time other system objectives are decreased, such as high passenger comfort or low energy consumption. During operation, the three objectives can be influenced by a variation of the controller parameters. The optimization parameters are the three Sky-Hook controller parameters in vertical, horizontal and rotational direction.

The main cause of hydraulic actuator failures is due to wear [127]. When in motion, the oil flow passes through the valve, leading to residue build up [112]. Since the system behavior is highly dynamic during normal operation, the valves are also subject to thermal strain due to dissipated electrical energy. This heating effect in turn influences the residue buildup and increases the probability of failures due to varnish. In order to increase the dependability of the valves, both the dissipated energy as well as the valve motion have to be minimized. However, if the valve were to be simply shut down and left at rest, the possibility of the valve becoming stuck permanently would rise unacceptably [111].

All these effects of hydraulic valve deterioration correspond to either the oil flow, which depends on the position of the spool, or the electric energy dissipated in the coil of the valve, which corresponds to the position and velocity of the spool. The position of the spool corresponds to the potential energy $E_{P,j}(t)$ stored in the valve spool return spring (stiffness c_v) while the spool velocity corresponds to the kinetic energy $E_{K,j}(t)$ stored in the spool (moving mass m_v). To include all effects of hydraulic valve deterioration, the kinetic and potential energy of the jth valve are weighted separately using a weighting function $W_j(E_{P,j}, E_{K,j})$:

$$W_j(E_{P,j}, E_{K,j}) = a(d + E_{P,j}) - b + \frac{1}{c(e + E_{K,j})(d + E_{P,j})}.$$

The function has been parameterized to give the desired properties ($a = 10/9$, $b = 19/81, c = 2916000/361, d = 1/180, e = 1/100$). If the valve is open ($E_{P,j} \neq 0$), thus leading to oil flow and actuator motion, the value of the weighting function rises almost linearly with the potential Energy $E_{P,j}$. If the valve is at rest ($E_{P,j} = 0$, $E_{K,j} = 0$), a finite result $W_j = 2$ is obtained, thus penalizing very low valve motions.

The potential and the kinetic energy are given by:

$$E_{P,j}(t) = \frac{1}{2} \cdot c_v \cdot x_j(t)^2,$$

$$E_{K,j}(t) = \frac{1}{2} \cdot m_v \cdot v_j(t)^2.$$

To obtain the spool's position $x_j(t)$ and velocity $v_j(t)$, an observer is used for the valve.

The signal is then averaged for one full simulation run ending at time T:

$$W_{a,j} = \frac{1}{T} \cdot \int_0^T W_j(E_{P,j}(\tau), E_{K,j}(\tau))\,d\tau.$$

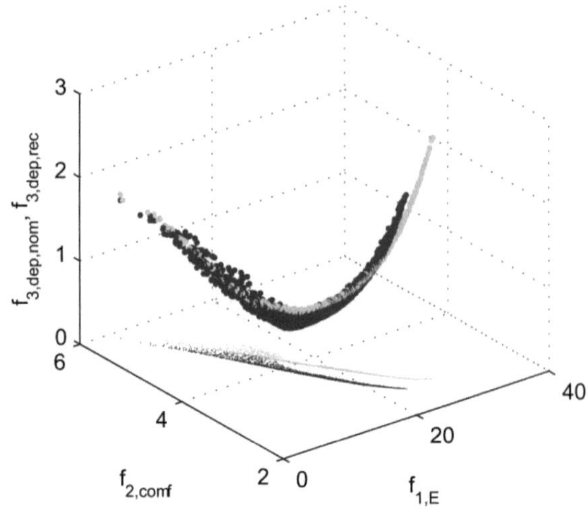

Fig. 3.13 Objectives in the nominal case (light gray) and objectives of the reconfigured system (black). The three objectives depicted here are: $f_{1,E}$: required power; $f_{2,comf}$: passenger *dis*comfort; $f_{3,dep,nom/rec}$: *un*dependability in the nominal and in the reconfigured case.

The final dependability value for the nominal case is calculated assuming that the actuator's signals are normally distributed and by calculating mean μ and standard deviation σ:

$$f_{3,dep,nom} = \mu\left(p_\mu \cdot \left[W_{avg,1},\ldots,W_{avg,6}\right]\right) + \sigma\left(p_\sigma \cdot \left[W_{avg,1},\ldots,W_{avg,6}\right]\right)$$

The weighting factors p_μ and p_σ have been parameterized empirically [101].

When the system is classified as Level II, the results of the optimization with the additional objective "minimize *un*dependability" are used to influence the system behavior. The objectives are depicted in Fig. 3.13 (light gray). The shape of the Pareto optimal points is restricted almost entirely to a one-dimensional line. An optimal parameter set of the Sky-Hook controllers pertains to each point. The original behavior regarding the two objectives "minimize energy" and "minimize *dis*comfort" is maintained as expected. The objective "minimize *un*dependability" is plotted on the Z-axis. Note that $f_{3,dep}$ depends on the system state, giving rise to two individual functions: $f_{3,dep,nom}$ for the nominal case with six actuators and $f_{3,dep,rec}$ for the reconfigured case with less than six actuators. It is obvious that the two objective functions "minimize energy" and "minimize *un*dependability" are in the same direction over a wide range. The more energy consumed by displacing the hydraulic cylinders, the less dependable the system is. At the other end of the curve, the opposite effect occurs. With less cylinder motion, the probability rises that the valve will become stuck. To obtain dependability-optimal operation, which corresponds to f_3 "minimize *un*dependability" being minimal, it is necessary to use small actuator motions with low dynamics without bringing the actuators to a stop.

In the case of falling below a certain threshold of remaining useful life, the system deterioration state is classified as Level III. As a result, the vulnerable actuator is shut down and the control of the remaining actuators has to be reconfigured to take the altered system structure into account. Depending on the position of the actuators, up to three actuators can be shut down.

Many approaches have been developed for the reconfiguration of an entire control system. In our work, a method based on the system description with a linear state space equation is applied to reconfigure the system [22] . This linear control reconfiguration method was originally developed for systems subject to actuator failures; nevertheless, it can be used for an intended actuator shutdown as well. A reconfiguration block is integrated between the altered system and the nominal controller. It modifies the control inputs of the remaining actuators and thus compensates for the effects of the deactivated actuator on the dynamical behavior. The main advantage of this method is the retention of the nominal controller in the closed-loop system. In this example, the five remaining cylinders have to perform the task of the nominal six cylinders.

At Level III of the M-LD Concept, it is still possible to apply the aforementioned multiobjective optimization even though the system structure has been reconfigured. If, for example, only one cylinder is switched off intentionally, the three objectives and the optimization parameters still remain, except that the vulnerable cylinder is not taken into account while calculating the energy- and dependability-objective functions. To achieve this, the objective "minimize *un*dependability" is altered accordingly.

The result of the multiobjective optimization of the reconfigured system with an intended actuator shutdown is shown in Fig. 3.13 (black). The shape of the set of the Pareto optimal points is still mostly the same as in the fully operational case. However, for a given level of *un*dependability, the reconfigured system achieves the same level of comfort for each Pareto optimal point, with less energy consumption than the non-reconfigured system.

The adaption in Level II is carried out by modifying the priority level of the individual objectives using the set of objective functions f_1, f_2 and $f_{3,nom}$. Since certain damping parameters are tied to each Pareto optimal combination of the three objectives' priorities, the system behavior is adapted by setting the parameters accordingly. An increased priority for the objective "minimize *un*dependability" leads to beneficial system behavior. However, one of the other objectives will be affected negatively, leading to either increased body motions or increased energy consumption.

For Level III, an intended actuator shutdown is initiated. Upon initiation, the reconfiguration block is activated in the model of the dynamical behavior and, in addition, the objective functions used for the multiobjective optimization are changed as well. During operation, a different set of results from prior optimization calculations is selected in order to adapt the system behavior.

3.2.1.5 Further Reading

The application example is explained in more detail in [101]. Another example of an application of this method is shown in [82].

Several additional methods have been developed to support the development process. The M-LD Concept can be implemented using information that is available during early development phases; this method is explained in detail in Sect. 3.1.2. To determine suitable counter-measures, a close interaction with multiobjective optimization (see Sect. 3.2.11) is required.

3.2.2 *Iterative Learning of Stochastic Disturbance Profiles*

Martin Krüger and Ansgar Trächtler

This section deals with learning excitation data from a specific class from disturbances. The quality of information about the momentary situation is particularly important in determining the success of any self-optimization strategy. Model-based methods for estimation of system states and physical parameters have been developed and implemented and have proven to be effective for the application of classical control techniques, e.g. [93]. A self-optimizing system, however, requires additional information about its environment, e.g. about future excitations and disturbances. With this information, it is possible to develop a system that is able to react intelligently and results can be achieved that surpass those of classical control strategies.

In recent research the combination of model-based disturbance observers and iterative learning methods has proven to be effective. In [146], the RailCab, an autonomously driven railway vehicle [126], serves as an application example: the rail-track deflections in lateral and vertical directions were learned and later used for disturbance compensation and self-optimization tasks.

In this section, we transfer this approach to the task of learning road disturbance profiles. In contrast to rail-track deflections the excitations and disturbances of roads do not occur deterministically when driving over the same road section several times, due to possible changes of the lateral position of the vehicle. Hence, a purely model-based approach is no longer beneficial. Instead, we use a combination of a behavior-based model of the road excitations and a model-based disturbance observer. These road excitations are modeled as nodes in a Bayesian network and a method is implemented to learn both parameters and parts of the network structure. The main idea is to use stochastic dependencies between different excitations (see Fig. 3.14). Disturbance profiles can help to increase the safety of self-optimizing systems, as they can be used for planning tasks or specific controller adaptation.

The iterative learning approach is based on the Principle Solution. Several aspects of the domain-spanning description are used: the partial models "Application Scenarios" and "Environment" yield general information about disturbances and in which situations they occur, while the partial models "Active Structure" as well as "Environment" show how disturbances influence the system and which sensors are available for the disturbance observer.

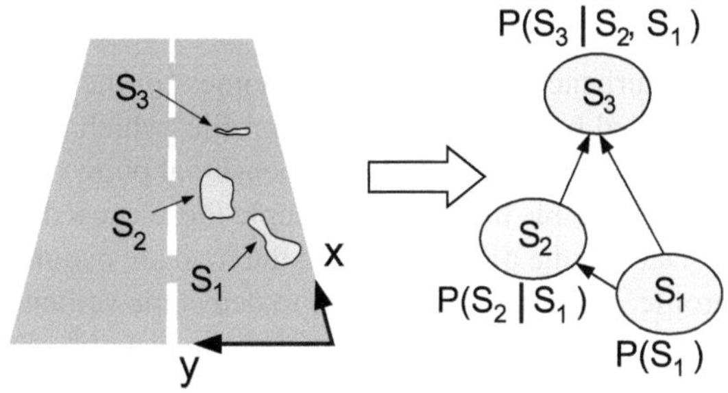

Fig. 3.14 Three Disturbances S_1, S_2 and S_3 on a sample road and corresponding Bayesian network

3.2.2.1 Prerequisites and Input

The presented approach is applicable to the disturbance estimation of road vehicles. It can also be generalised to other systems with stochastically dependent occurences of disturbances; however, we will only focus on road profiles at this stage.

In order to learn road profiles, several prerequisites have to be fulfilled in addition to the above-mentioned partial models of the Principle Solution. Firstly, data about road excitations are necessary. Mostly, the excitation cannot be measured directly, as sensors cannot be put between the wheel and the road surface. Therefore, we have developed a model-based disturbance observer to detect disturbances. A model of the vehicle as well as a disturbance model, are needed to design the observer. Additionally, sufficient empirical data, such as vertical accelerations of the body and wheel have to be available. By means of the disturbance observer, a current road profile of a particular road section can be computed based on the measured values. This constitutes the input for the learning method. We also assume that the current longitudinal position of the vehicle on the road is known.

3.2.2.2 Description

The developed learning method consists of three sequential steps (Fig. 3.15). To start, the disturbances due to road unevenness have to be detected and identified on the basis of different passages. Each disturbance is then set up as a node of the Bayesian network. Additionally the parameters, i. e. the conditional probability distributions for every single node, have to be acquired from the data. After compilation, the Bayesian network can be used to predict disturbances during following driving sequences over the same road section. A description of these three steps is presented in the following section.

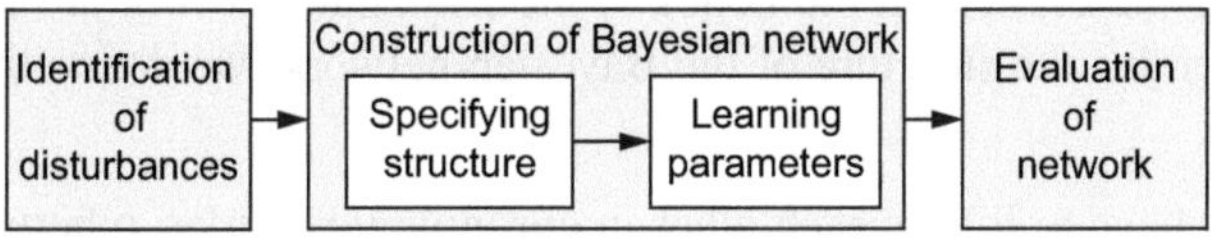

Fig. 3.15 Structure of the learning process

Identification of Disturbances

A disturbance is caused by unevenness in the road surface. These "defects" can be described by a discrete surface profile, which assigns a height (the position in vertical direction) to each two-dimensional point on the road. Disturbances are abrupt changes in this height, for example those caused by potholes.

Data on the disturbances on the road are acquired by means of a vertical velocity profile. In our case, it is easier to detect the velocity than the position, as an absolute road profile is hard to determine by means of acceleration-based sensor concepts. The derivative of the position yields peaks that can be used for a precise detection of relevant disturbances.

Due to inaccuracies in the measurements (e.g. sensor noise), both the position and the strength of the disturbance are slightly different for every measurement; hence, different clustering methods are used for the identification of disturbances. Clustering has to be employed because the total number of disturbances can only be detected using the data from different driving sequences. Using clustering the data can be suitably merged because it describes the combination of similar objects in groups.

The actual position of the disturbances is determined by a hierarchical clustering of the positions detected on different passages [4]. In this context, "hierarchical" means that several levels of clusters are created in consecutive steps. On every hierarchical level, the elements of the previous level with the shortest distance between them are merged into new clusters, thus yielding a binary tree structure with the single elements as leaves. This hierarchical clustering produces very good results for our application, as shown in Sect. 3.2.2.4.

In the second step, the strength of each disturbance is determined by clustering with Gaussian mixture distributions [4]. This method has been applied because, presumably, the strength of the disturbance (vertical direction) during different passages is distributed normally due to measurement errors or irregular obstacles. In our approach, the density estimation of each expected disturbance is reduced to the determination of the expected value and the standard deviation of the Gaussian distribution. With the help of the *Expectation-Maximization* algorithm [4] a mixed distribution can be found which best fits the distribution of the disturbances, i. e. it describes the data with a high degree of probability.

With the results of the cluster analysis, which we carried out using the methods described, the expected disturbances and thus the nodes of the Bayesian network can be determined, i. e. each cluster results in one node as a detected disturbance.

Construction of Bayesian Network

As a result of the clustering, the nodes of the Bayesian network are identified. In order to determine the structure of the Bayesian network, we define the following rules:

- The nodes have to follow each other in chronological order; otherwise a causal relation makes no sense.

- Two disturbances located next to each other in lateral direction cannot be crossed one after another. Therefore, no edge is allowed between such nodes.
- A stochastic dependence of nodes that lie far apart from one another is improbable, so there should be no edge between nodes with a distance between them exceeding a pre-determined value.

These rules yield the complete structure of the Bayesian network.

The parameters are determined by means of the *Maximum-Likelihood* algorithm on the basis of measured data. So the construction of the Bayesian network is completed.

Evaluation of the Network

After construction, the Bayesian network can be used for predicting disturbances during later driving sequences. On one hand, we can determine the a priori probability, i. e. the unconditional probability, of the occurrence of every disturbance. On the other hand, we can obtain more detailed information from an input of information, so-called evidence, into the network. By evaluating random variables (evidence), i. e. the information of wether single disturbances were driven over or not, we can evaluate the conditional probability distributions by drawing conclusions within the network, so-called inference. These calculated probabilities, along with the saved data in the nodes, allow a prediction of the position and strength of the following disturbances.

3.2.2.3 Results

The result of the described iterative learning method is a Bayesian network that represents road disturbances for a particular road segment. The nodes provide information about position and strength of the disturbances in a compact format. But the main benefit is given by evaluation of the probabilities inside the Bayesian network. The unconditional probailities can be used to classify roads in general. In addition, upcoming disturbances can be predicted by means of the conditional probabilities.

3.2.2.4 Application Example

The iterative learning process has been tested and validated using a simplified model of the X-by-wire test vehicle "Chameleon" , introduced in Sect. 1.3.3. We have here only considered the vertical dynamics. As usual, we use a quarter vehicle model because it provides the essential aspects for a first verification of the iterative learning approach. This leads to the model topology presented in Fig. 3.16.

In contrast to common quarter-vehicle models, we have more than just two degrees of freedom, z_B and z_W, for the body mass and the wheel mass respectively. Because the driving motor with its non-negligible mass is used as a mass absorber, it must be accounted for in the model by an additional degree of freedom z_D. The elasticity between the driving motor and the wheel is modeled by a simple generalized Maxwell element representing the real elastomer mounting. The vertical

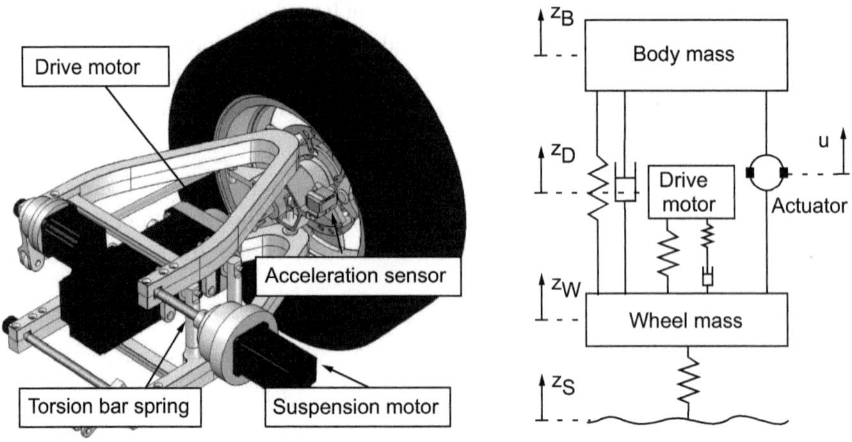

Fig. 3.16 Corner module of the test vehicle *(left)* and physical surrogate model of vertical dynamics *(right)*

dynamics can be controlled by forces between the body and the wheel, which are applied by the motor of the active suspension.

Building the simulation model now comprises two steps. Firstly, we need an appropriate model of the dynamics to simulate the effects of road disturbances. Additionally, we have to design a disturbance observer to compute the input data for the iterative learning process.

The dynamics of the system shown in Fig. 3.16 can be represented by a linear system with seven states, i.e. $x \in \mathbb{R}^7$, one control input u and one disturbance input z. The model has two outputs $y \in \mathbb{R}^2$ which are composed of both the wheel and the body acceleration and conform to the real sensor concept. With matrices of appropriate size we get the linear state-space representation

$$\begin{aligned} \dot{x} &= Ax + Bu + Ez, \\ y &= Cx + Du + Fz. \end{aligned} \tag{3.3}$$

The observer model will not represent the real behavior exactly. To consider this effect in our simulations, we use parameters for generating the vehicle model that vary slightly from the parameters used in the model of the disturbance observer.

We use a simple first order transfer function G_M to model the dynamics of the acceleration sensors combined with an additional Gaussian noise signal. The controller has been designed as an optimal Linear-Quadratic (LQ) controller [106].

The resulting simulation model with all relevant components has the structure shown in Fig. 3.17. It can be seen that the disturbance observer has to fulfill two tasks. On one hand, it has to estimate the state variables for use by the controller, and on the other hand, it has to detect the disturbance $\hat{z}$ and estimate the derivative $\hat{z}'$ for the learning process.

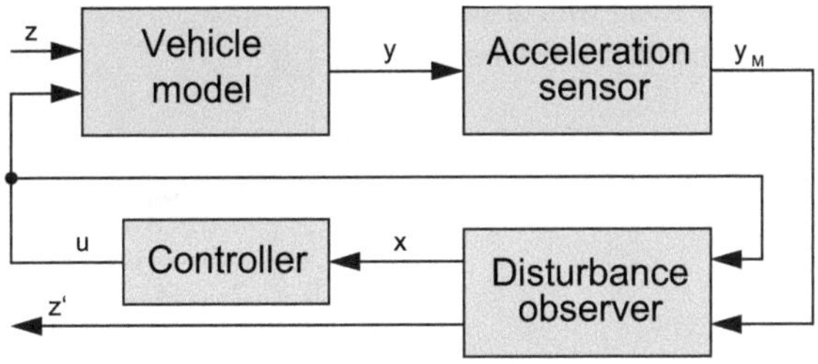

Fig. 3.17 Structure of the simulation model

An estimation of the disturbances can be obtained by adding a disturbance model to the standard state observer. This disturbance model is the linear model

$$\begin{aligned} \dot{x}_S &= A_S x_S, \\ \left[\hat{z}\ \hat{z}'\right]^T &= \left[C_S\ C_Z\right]^T x_S \end{aligned} \tag{3.4}$$

with two outputs $\hat{z}$ and $\hat{z}'$. $\hat{z}$ is needed by the observer itself to consider the influence of the disturbance on the vehicle dynamics. $\hat{z}'$ can be used directly by the learning process. The dynamics of the disturbances are characterized by additional states x_S and a disturbance dynamic matrix A_S. For our example, we have chosen a two-dimensional disturbance model and have assumed that the second derivative of the disturbance is constant, which leads to a singular matrix A_S.

The complete observer is implemented as a continuous-time Kalman filter [106]. It combines the quarter vehicle-model (Eqn. 3.3) and the disturbance dynamics in one dynamical system

$$\begin{aligned} \begin{bmatrix} \dot{\hat{x}} \\ \dot{x}_S \end{bmatrix} &= \begin{bmatrix} A & EC_S \\ 0 & A_S \end{bmatrix} \begin{bmatrix} \hat{x} \\ x_S \end{bmatrix} + \begin{bmatrix} B \\ 0 \end{bmatrix} u + L_S (y_M - \hat{y}), \\ \begin{bmatrix} \hat{y} \\ \hat{z}' \end{bmatrix} &= \begin{bmatrix} C & FC_S \\ 0 & C_Z \end{bmatrix} \begin{bmatrix} \hat{x} \\ x_S \end{bmatrix} + \begin{bmatrix} D \\ 0 \end{bmatrix} u. \end{aligned} \tag{3.5}$$

The observer matrix $L_S \in \mathbb{R}^{9\times 2}$ defines the dynamics of the observer. It is computed using the standard techniques for designing Kalman filters, i.e. by defining two covariance matrices describing the noise of the system and computing the optimal matrix L_S. In our example, the covariance matrices are chosen so as to yield best results for the estimation of $\hat{z}'$. In specific terms, this means that the variances corresponding to x_S are higher than those corresponding to the remaining ones.

For testing purposes, excitation data for the quarter-vehicle model were generated in three steps:

- A simple road model was built, representing the road surface by a plane divided into discrete sections. Road disturbances, such as potholes, were defined by assigning a height value to each section. For the tests presented below, a number of simple disturbances have been used to generate excitation data.
- In the next step, 20 straight trajectories along the road were created randomly.

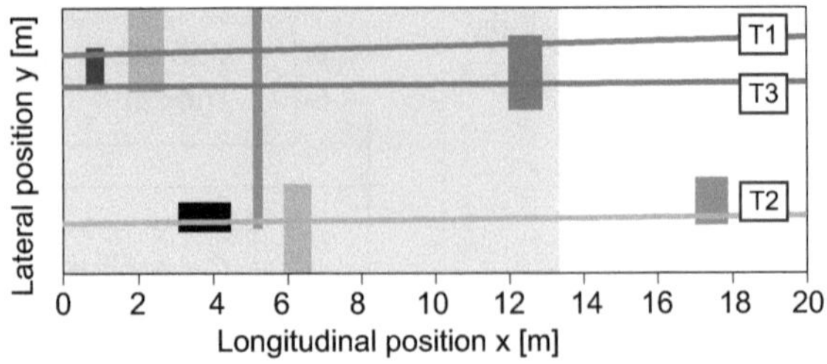

Fig. 3.18 Road with disturbances and three trajectories

- The height profile of the road along the trajectories defines the excitation given as input for the vehicle model.

With these steps, road profiles of 20 trajectories have been generated. Their data were used as excitations for the quarter-vehicle model, simulating 20 driving sequences along the road. During each iteration, the excitation was estimated by the disturbance observer. The occurring disturbances were learned by providing the learning algorithm with data from the observed excitation and with the longitudinal position of the vehicle.

To verify the results, we used the constructed Bayesian network to predict disturbances during three other simulated passages on random trajectories. During those passages, probabilities of the occurrence of learned disturbances were calculated and constantly updated. These probabilities were calculated using the Bayesian network and were based on evidence given by disturbances already struck or passed during the current driving sequence.

Figures 3.18 and 3.19 illustrate the results of the test runs. Figure 3.18 shows the road model with the disturbances that were learned. The three lines represent the trajectories used for the prediction test. The driving direction of the modeled quarter-vehicle is from left to right.

The excitations ($\hat{z}'$) observed by the model are shown in Fig. 3.19 in relation to the longitudinal position x. The disturbances are visible as peaks. The circles, squares and crosses mark the position and intensity of learned disturbances. The numbers indicate the calculated probability of hitting the corresponding disturbance, obtained by inference in the Bayesian network. The plotted values are calculated online during the driving sequence considering all evidence obtained up to that point. The crosses mark the position of learned disturbances which are assumed to be hit according to their probability. The circles show disturbances with low probabilities, which are considered as not being hit during that particular iteration.

It can be seen in the figures that most predictions based on the calculated probabilities are correct, which means that the disturbances are considered correctly as either being missed (circles) or being hit (crosses). As it turns out, of 25 disturbances hit during the three runs shown in Fig. 3.19, 21 have been predicted correctly, which is more than 80 %. Furthermore, the forecasts for all disturbances that were not hit were accurate. Altogether, the occurrence or non-occurrence of about 90 % of the learned disturbances could be predicted correctly.

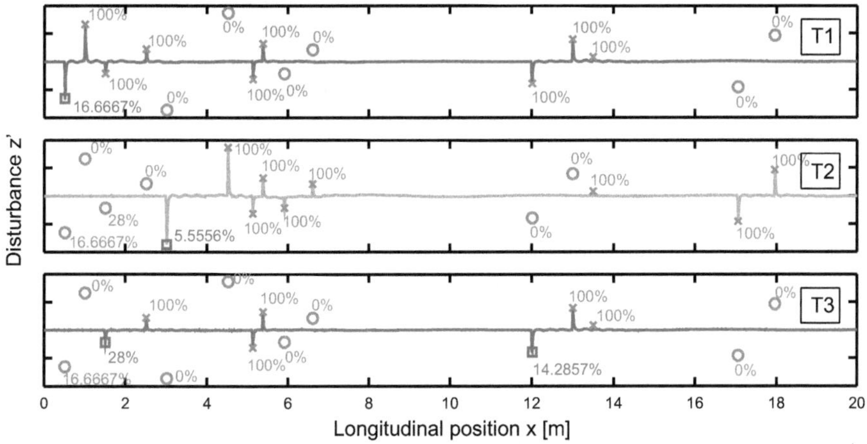

Fig. 3.19 Derivative of excitations (trajectories of Fig. 3.18) and results of prediction. (*Circles*: correct prediction of non-occurence, *Cross*: correct prediction of occurence, *Squares*: false prediction (false-negative))

Summing up the results, we can state that the probabilistic iterative learning method based on Bayesian networks can be successfully used to learn and predict randomly occurring, stochastically dependent excitations in the context of road vehicles. The learned data could conceivably be used in different ways. In the context of self-optimizing systems, applications such as classification of road segments or planning tasks become possible. The method can also be beneficial for the system safety, as it offers additional information about the environment; e. g. dependability-oriented configurations could be used by the system for road segments with especially uneven surfaces.

3.2.2.5 Further Reading

A more detailed describtion of the method can be found in [19].

3.2.3 Mutation Testing of Electronic Component Design

Wolfgang Müller and Tao Xie

The complexity of self-optimizing systems requires systematic and thorough verification of their designs, in order to guarantee adaptive yet desired runtime behavior. In this section, we focus on the design verification of electronic components, such as an embedded microprocessor, and propose an adaptive verification method that is directed by coverage metrics.

Together with other electronic peripherals, microprocessors comprise the central platform hosting hardware dependent software and application software, which is the seat of system intelligence. The design of these components must be verified as comprehensively as possible, as their later integration into a complex system makes in-system verification difficult and their functional correctness should be ensured beforehand. We assume that simulation remains the primary platform for functional verification and prototyping of electronic designs, though other alternatives do exist, such as the model-checking used in Section 3.2.7.

In this context, we can find coverage metrics at the heart of verification thoroughness. With coverage, we systematically and quantitatively measure the progress of a simulation-based verification process. Various code coverage and toggle coverage are the primary metrics. Recent research extends to more complex functional coverage [50] and assertion-based coverage.

Mutation testing, also called mutation analysis, is a unique coverage metric that assesses the quality of test data in a more stringent manner. It was originally introduced and studied as a theory for software testing [36]. Later, mutation testing was increasingly considered for hardware design verification [136] and an industrial EDA (Electronic Design Automation) tool for Hardware Description Languages (HDLs) mutation testing became available [67].

We have chosen mutation analysis as the advanced, representable coverage metric most suitable for our purposes. In the following section, we will first introduce the basic principle of mutation analysis and then present a mutation analysis feedback-directed adaptive testbench for functional verification of electronic component designs. The method has been evaluated using a typical embedded microprocessor design.

3.2.3.1 Prerequisites and Input

Mutation testing is a fault-based simulation metric. It highlights an intrinsic requirement for simulation test data that they should be capable of stimulating potential design errors and propagating the erroneous behavior to checking points. Mutation testing measures and enhances the simulation process as shown by Fig. 3.20.

A so-called mutation is a single fault injection into the code of a copy of the Design Under Verification (DUV), such as this HDL statement modification:

$$\text{a} <= \text{b and c;} \xrightarrow{mutation} \text{a} <= \text{b or c;}$$

The fault-injected copy is called a mutant of the design. For each test case, the mutant is simulated after the simulation of the original design and the results of both simulations are compared. If any difference appears at the output during the simulation, this test is said to be able to "kill" the mutant. Each type of fault injection is called a mutation operator and dozens of such operators can be defined based on the applied design language. By deploying these pre-defined mutation operators at different locations of a design, we can obtain a huge database of mutants. Then the number of killed mutants determines a mutation coverage metric, which can be used to measure the overall quality and thoroughness of the testbench and the entire simulation process.

We consider random simulation a long-recognized useful lightweight method of carrying out mutation testing. However, the lightweight nature of random simulation

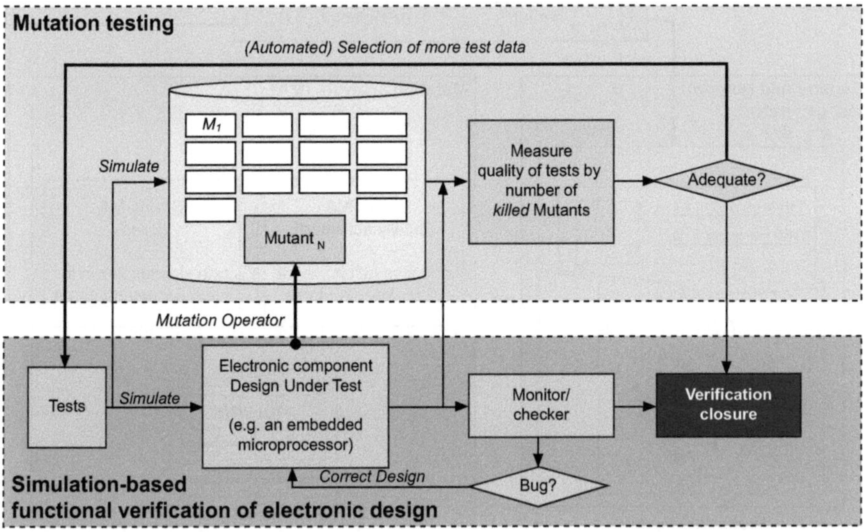

Fig. 3.20 Mutation testing for the functional verification of electronic component designs

is in contrast with the inherent computation expensiveness of mutation testing. In essence, each time a random test is generated, it should be simulated against not only the original Design Under Verification, but also against all the mutants that have been created as the cover points, of which there is a massive amount. Since the test is randomly selected and relatively undirected, this amplifies the mutation testing problem.

3.2.3.2 Description

Therefore, we propose a mutation testing feedback-directed adaptive random simulation method. Our proposal improves the efficiency of the mutation testing coverage and solves the inherent conflict between random test generation and mutation analysis. We propose the use of a constrained Markov chain to enable effective adjustment to the probability model of random simulation. An efficiency-improving heuristic makes this adjustment by utilizing two-phase mutation testing results. Such a testbench is shown in Fig. 3.21. The idea is to integrate an in-loop heuristic process that dynamically adapts the test probability model to a more efficient distribution for mutation coverage. On-line results from each mutation-testing run are analyzed to derive the adjustment. Our evaluations demonstrate that the heuristic reaches a higher mutation coverage in less simulation time.

As a prerequisite for the dynamic adjustment, a probability model of test sequences is required that provides the possibility of parameter steering. Considering that an electronic component design has a precisely defined instruction interface, such as the ISA (Instruction Set Architecture) of a microprocessor, or the communication protocol of a bus controller, test inputs in a random test generator are modeled

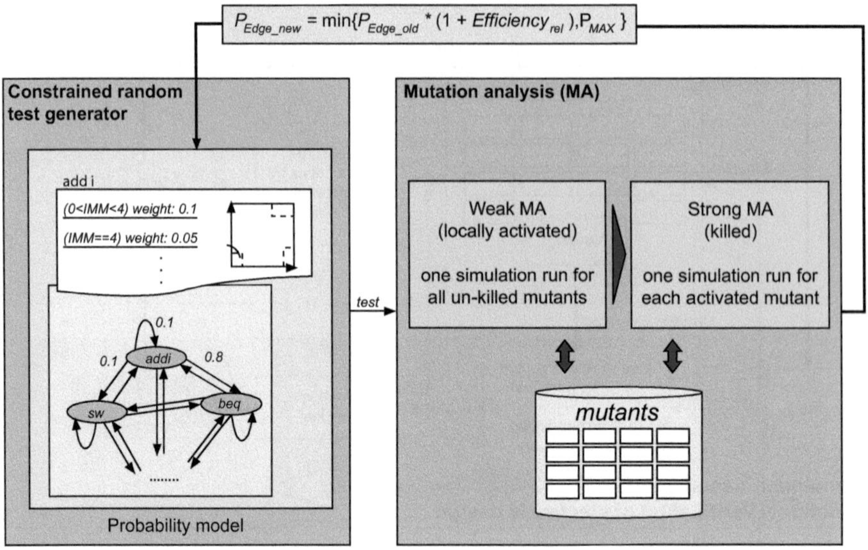

Fig. 3.21 A mutation testing directed adaptive simulation framework for the functional verification of electronic component designs

in two layers as shown in Fig. 3.21. In a first step, a Markov chain is used to represent sequences of tests. Each node models one type of test instruction. The selection probability on edges enables us to establish the correlation between mutation analysis efficiency and a short pattern of test sequences. Second, weighted constraints are defined on the fields of an instruction. This provides the possibility of steering test patterns towards more effective areas, such as corner cases.

Each time a test is generated, we record the pair of Markov edge and constraint selected for the generation. The basic idea is to estimate the efficiency of each test on mutation analysis and use the estimation to adjust the probability of the corresponding Markov edge and constraint. This efficiency estimation should follow the unique simulation process of mutation analysis. As the right half of Fig. 3.21 shows, we first introduce an extra weak mutation analysis [72] phase. It uses one simulation cycle to identify the locally activated mutants. Only those are fed into a traditional, strong mutation analysis phase and are fully simulated, so that we can see whether they are killed using the criterion that a different value appears at design output ports. Consider that φ is the test probability distribution from a Markov chain/constraint model, which further implies $P_{M_i\_activated}$ and $P_{M_i\_kill}$ for each mutant M_i as its probabilities of being activated and killed using the current test model. On a set of N_{Mutant} design mutants, this leads to an *expected simulation effort* for the mutation analysis flow in Fig. 3.21, represented as

$$\max_{1 \leq i \leq N_{mutant}} (1/P_{M_i\_kill}) + \sum_{1 \leq i \leq N_{mutant}} (P_{M_i\_activate}/P_{M_i\_kill})$$

Based on this expected simulation effort, we use the number of mutants activated by the test $N_{activated}$ and the number of mutants killed N_{killed} to estimate the efficiency of this test as

$$Efficiency = \frac{N_{killed}}{N_{activated}}$$

A low ratio means that too many mutants are merely activated and that a large number of simulations are wasted in the second phase without killing the mutants. We also record this efficiency value for the last 10 tests generated and use the average $Efficiency_{average\_last\_ten}$ to derive a relative value that lies between 0 and 1

$$Efficiency_{rel} = \frac{Efficiency}{Efficiency_{average\_last\_ten} + Efficiency}$$

According to this, at the early stage of a random simulation, test patterns with high mutation kill/activation rates are encouraged. However, we observed in our experiment that in the last stage, it may well happen that no single mutant is killed in ten consecutive iterations. In such a case, the heuristic approach *changes to another mode that encourages more activation of mutants*, by first calculating *efficiency* as an adjustment value and then increasing the probability/weight of the corresponding Markov chain edge/constraint with the following value:

$$Efficiency_{rel} = \frac{N_{activated}}{N_{activated\_average\_last\_ten} + N_{activated}}$$

It is safe to assume that there will always be some mutants activated.

Initially, all Markov chain edges have the same probability of being selected and instruction constraints have the same weight. At the end of each iteration for test generation, the probability of the edge used as well as the weight of the constraint used is adjusted by

$$\begin{cases} P_{Edge\_new} = \min\{P_{Edge_old} * (1 + Efficiency_{rel}), P_{MAX}\} \\ P_{Constr\_new} = \min\{W_{Constr_old} * (1 + Efficiency_{rel}), W_{MAX}\} \end{cases}$$

P_{MAX} and W_{MAX} are efforts to prevent the starvation of other edges/constraints by setting an upper bound of probability on one edge/constraint. In a real-world application that follows, these two numbers were set to 0.9 for a model of 58 Markov edges, for instance.

For each $Edge_i$ that branches out from the same instruction node and each $Constr_i$ on this node, we adjust their probability/weight proportionally to their old values

$$\begin{cases} P_{Edge_i\_new} = (1 - P_{Edge\_new}) * \dfrac{P_{Edge_i\_old}}{1 - P_{Edge_i\_old}} \\ P_{Constr_i\_new} = (1 - P_{Constr\_new}) * \dfrac{P_{Constr_i\_old}}{1 - P_{Constr_i\_old}} \end{cases}$$

As a heuristic approach, the formulation of the adjustment is motivated by two points.

First, the ultimate problem of test generation in mutation analysis is to kill a design mutant, the mutant simulation is required to first *reach* the mutation statement, then *activate* this mutant by executing the mutated statement in such a manner that a local deviation is created, and *propagate* this deviation to any output of the design.

Second, there are two hypotheses that we consider reasonable for correlating mutation analysis feedback to the test generation problem: *activation-kill* and *similar-test.*

The *activation-kill hypothesis* states that *if a test activates a large number of mutants in a simulation, it also leads to a simulation that kills many mutants in the end.* In other words, the *mutant-activation efficiency* of a test is coupled with its final *mutant-killing efficiency.* This is actually straightforward based on the problem of mutation analysis test generation – activation is a necessity condition for killing mutants, preceding propagation.

The *similar-test hypothesis* states that *if a test activates a lot of mutants, the pair of Markov chain edge/constraint that was used for generating this test will further generate tests that activate similarly many mutants.* In short, a pair of Markov chain edge/constraint represents a type of test. We expect that tests generated from the same type have *similar* mutant-activation efficiency.

This can be taken as an explanation for why the heuristic adaptation approach is described as it is in this section.

3.2.3.3 Results

The result of our adaptive verification method, when applied to an electronic component design, is a random simulation testbench that is able to achieve a higher mutation coverage with less simulation effort.

3.2.3.4 Application Example

We have created an experimental version of the mutation analysis directed simulation method and applied it to the functional verification of the MB-LITE microprocessor design [85]. This soft microprocessor core realizes MicroBlaze ISA in VHDL and is a widely used architecture in embedded systems. It is able to execute binary code compiled with the standard MicroBlaze compiler *mb-gcc* (included in the XILINX FPGA tool). Mutation testing was implemented by *Certitude*(TM), an industrial EDA tool, which selectively generated 732 mutants for the MB-LITE design.

The MicroBlaze instructions [153] were modeled by 58 Markov-chain nodes. Similar instructions are not considered distinctly, such as *add*, *addc*, *addk* and *addkc*. Further, 17 constraints were defined to partition instruction areas. We used the *SystemC Verification Library* for implementing the Markov chain and the associated instruction constraints, as it provides a convenient framework for probability modeling and constraint solving. SystemC and VHDL co-simulation is also supported by the *ModelSim* simulator. All the edges and constraints have an equal probability of being selected at the beginning.

We compared the following test generation processes: i) the random test generation under the constrained Markov-chain probability model and dynamically directed by the in-loop heuristics, ii) the same random test generation, but without the in-loop adjustment, and iii) two software programs: a basic "hello world" and a *Dhrystone* benchmark. These programs can be seen as competitive comparisons, since they are generated by the compiler's extensive knowledge of the instruction set.

Figure 3.22 shows their efficiency in terms of mutation analysis coverage, i.e. number of killed mutants versus test instructions used. The simulations lasted for several hours on a computer with a 2.4 Ghz processor. The data are drawn only to the points after which there was no further change in coverage.

The mutation analysis feedback-directed random simulation reached a status of killing 691 mutants within about 8.3 hours, compared to the non-adaptive variant that killed 625 mutants in 11 hours. The coverage is 94.4% versus 85.4%. The improvement in performance came firstly from the ability of the heuristics to steer the test generation towards more effective patterns in the early phase. Secondly, this efficiency is also amplified continuously in mutation analysis, as the early killing of mutants eliminated the cost of their simulation afterwards. Further, the heuristics encouraged more activation in the late phase, which showed the effect of trying-and-killing more mutants.

Both random simulations outperformed the other two software binaries. The basic "hello world" was able to kill 507 mutants and the *Dhrystone* able to kill 565 mutants. At the initial stage, they displayed an identical coverage curve. In the middle period, both had a long sequence of wasted cycles without contributing any coverage improvement. This is inferior to the continuous increase of killed mutants in random simulations.

Fig. 3.22 Efficiency improvement with adaptive random simulation, by comparing four simulation processes with regard to mutation testing

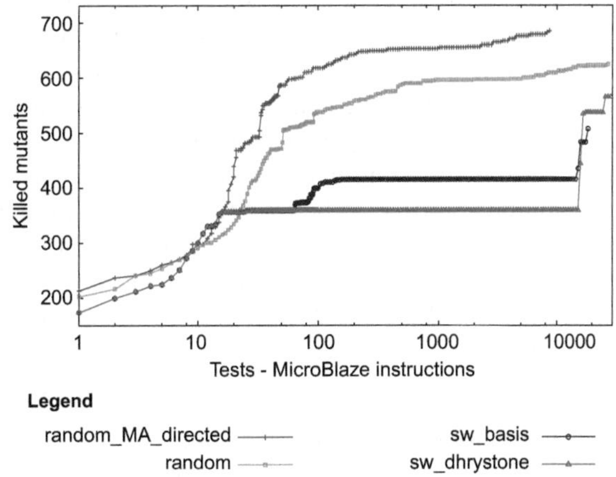

In the end, we came to the conclusion that the proposed adaptive random simulation method is able to substantially improve the testbench efficiency of targeting mutation testing and functions admirably as an advanced quality metric for the functional simulation of electronic component design in our self-optimizing systems.

3.2.4 Optimal Control with Uncertainty

Sina Ober-Blöbaum and Albert Seifried

The goal in classical optimal control theory is to determine control strategies that, if applied to the system, fulfill some predefined task optimally with respect to some given quantity. However, for many applications, specific system parameters such as e.g. friction coefficients or geometry parameters are unknown or cannot be measured exactly. Due to this uncertainty, the correct execution of the desired task can no longer be guaranteed by the control system. In this section, strategies for the optimal control of technical systems with uncertainties are presented under the aspect of the attributes safety and reliability by introducing a performance measure.

While in the deterministic setting, we characterize the maximal performance of the control system in fulfilling a predefined task exactly (e.g. a robot arm has to grab some tool in a specific position); in the presence of uncertainty we introduce a performance measure whose mean has to be minimized in order to guarantee maximal performance by the control system and thus lead to reliable system behavior. Regarding this notion of system performance as an additional objective leads to a multiobjective optimal control problem. Based on numerical solution methods for optimal control [107] and multiobjective optimization problems [34, 135], a numerical method is presented in this section to approximate the so-called Pareto optimal solutions of this multiobjective optimal control problem. The approach is verified by means of a robot arm maneuver for which the arm lengths are assumed to be uncertain.

For the application of the presented method, input from different partial models of the Principle Solution (cf. Sect. 2.1) is required. Concerning the control engineering methods, all information about the models of the dynamics have to be provided in order to formulate the control problem. The partial models "Application Scenarios" and "System of Objectives" provide all the necessary information for the task to be fulfilled by the control system, as well as the information with respect to which criterion this task has to be optimized. Furthermore, the partial models "Requirements" and "Environment" provide important information about uncertain model parameters and their ranges for the optimal control problem.

Let us consider a technical system with uncertainty given by the differential equation

$$\dot{\mathbf{x}}(t) = \mathbf{f}(\mathbf{x}(t), \mathbf{u}(t), \xi) \tag{3.6}$$

with Lipschitz continuous $\mathbf{f}$, the state function $\mathbf{x} : I \to \mathbb{R}^n$, the control function $\mathbf{u} : I \to \mathbb{R}^m$, $I = [t_0, t_f]$, and random variables $\xi \in \mathbb{R}^s$ with known distributions associated with the uncertainty of system parameters. It is assumed that $\mathbf{x}(t)$ can be expressed as a function of ξ, i.e. $\mathbf{x}(t,\xi)$.[4] Let $J(\mathbf{x},\mathbf{u}) = \int_I C(\mathbf{x}(t),\mathbf{u}(t))\,dt$ be a cost functional which measures a quantity of interest with the continuously differentiable cost function C. Then the expected value of J is given by

$$\hat{J}(\mathbf{u}) = \int J(\mathbf{x},\mathbf{u})\rho(\xi)\,d\xi, \tag{3.7}$$

where $\rho(\xi)$ is the probability density function (pdf) of ξ. A general optimal control problem is to find a control $\mathbf{u}^*$ that minimizes (3.7), subject to the differential equation (3.6). Common solution methods for solving these single optimal control problems are e.g. sample average approximation methods [25, 79], for which the expected value function is approximated by some sample average function based on Monte Carlo simulations. The resulting sample average optimization problem can be solved by standard optimization techniques (see e.g. [98]).

Typically, the intent is to design an optimal control such that the fulfillment of a task is guaranteed by the system (expressed by the terminal constraint $\mathbf{r}(\mathbf{x}(t_f)) = 0$). In this case, we say that system performance is maximal. Note that in the presence of uncertainty, the value of the terminal constraint $\mathbf{r}(\mathbf{x}(t_f,\xi))$ will be different for different values of the random variables ξ. Thus, the execution of the task can no longer be guaranteed by the control system. Instead, the goal is to design a control that, on one hand, is still optimal with respect to the prescribed objective functional and, on the other hand, optimizes the system performance, i.e. the system should fulfill the desired task as well as possible. To this end, we define a single performance measure $Y(\mathbf{x},\xi) \geq 0$ that depends on the random variables and which is, in the deterministic case, zero for maximal performance, i.e. $Y = 0$ if $r(\mathbf{x}(t_f)) = 0$. As measure of performance, we choose $Y = \|\mathbf{r}(\mathbf{x}(t_f))\|_2^2$. In the presence of uncertainty, the performance of the system is defined as optimal if the expected value of the performance measure Y is minimized, i.e. we want to minimize

$$\hat{J}_2 = \int Y(\mathbf{x},\xi)\rho(\xi)\,d\xi. \tag{3.8}$$

For the chosen measure of performance this means that maximal system performance corresponds to minimal mean violation of the terminal constraint.[5] This is in contrast to other approaches in robust optimal control, where, for example, the probability of a state constraint violation is required to be less than some small, but given probability, which is then included as an inequality constraint or as new single objective function in the optimal control problem (see e.g. [94]). Instead, Equation (3.8) is treated as an additional objective functional in the optimal control problem. By minimizing (3.7) and (3.8) subject to the differential equation (3.6), we are faced with a multiobjective optimal control problem, i.e. the optimization of not only one but several objectives at the same time is required. If the different

[4] In the following, we will simply write $\mathbf{x}(t)$ but assume that $\mathbf{x}$ is also a function of ξ.

[5] Note that the expected value is only one choice among many robustness measures and could be easily replaced.

objectives are conflicting, no unique optimum can be determined. Rather, we attempt to determine a set of compromise solutions: the Pareto optimal solutions [49].

3.2.4.1 Prerequisites and Input

Without loss of generality, a multiobjective optimization problem can be viewed as a minimization problem; however, in contrast to a (single) objective problem, the objective function is vector-valued (cf. Chapter 1.1.2.1).

Problem 3.1. A *multiobjective optimization problem* is given by

$$\min_{\mathbf{y}\in S}\{\mathbf{F}(\mathbf{y})\}, \quad S := \{\mathbf{y}\in\mathbb{R}^n \,|\, \mathbf{g}(\mathbf{y}) = 0, \mathbf{a}(\mathbf{y}) \leq 0\}, \tag{3.9}$$

with objective functions $\mathbf{F}:\mathbb{R}^n \to \mathbb{R}^k, k > 1$ and constraint functions $\mathbf{g}:\mathbb{R}^n \to \mathbb{R}^m$ and $\mathbf{a}:\mathbb{R}^n \to \mathbb{R}^q$.

For the computation of the set of Pareto optimal solutions (cf. Chapter 1.1.2.1) of Problem 3.1, there exists an extremely broad variety of different methods. The methods are classified as global or local methods, whereby the global methods iterate step-for-step a complete set of parameters to obtain an approximation of the Pareto set. Contrary to global methods, local methods use local information from single Pareto points to compute proximate solutions. Well-known global methods are evolutionary algorithms (e.g. [32]) or subdivision methods (e.g. [34]). Typical local methods are the weighted sum method (e.g. [32, 102]), continuation methods (e.g. [70, 131]) or reference point methods (e.g. [33, 102]). The latter methods are appropriate for approximating the Pareto set, especially for a high dimension n. The basic idea is to generate unreachable targets (or reference points) $\mathbf{T}\in\mathbb{R}^k$ in the image space of $\mathbf{F}$, where each of these targets is used for the following distance minimization:[6]

$$\min_{\mathbf{y}\in S} \|\mathbf{F}(\mathbf{y}) - \mathbf{T}\|, \quad S := \{\mathbf{y}\in\mathbb{R}^n \,|\, \mathbf{g}(\mathbf{y}) = 0, \mathbf{a}(\mathbf{y}) \leq 0\}. \tag{3.10}$$

Standard minimization algorithms like SQP [60], which is implemented in the NAG library[7] (*Numerical Algorithms Group*), or Ipopt[8] can be applied to solve these single-objective minimization problems. This yields a set P of minima which approximates the whole Pareto set if it is convex, and a continuous connected part of it in the nonconvex case.

To be more precise, for the given multiobjective optimization problem 3.1, a norm $\|\cdot\|$, and an optimal point $\mathbf{y}_0 \in \mathscr{P}$ the reference point optimization algorithm works as follows:

$P := \{\mathbf{y}_0\}$
$FP := \{\mathbf{F}(\mathbf{y}_0)\}$

[6] For the numerical computations, we minimize the squared distance to ensure differentiability.

[7] `http://www.nag.co.uk`

[8] `https://projects.coin-or.org/Ipopt`

for $i = 0,\dots,M$ **do**
  for $j = 1,\dots,k$ **do**
    Choose $\mathbf{T}^i_j \in \mathbb{R}^k$ near $\mathbf{F}(\mathbf{y}_i)$ but outside of $\mathbf{F}(S)$
    Solve $\mathbf{y}^*_j := \arg\min_{\mathbf{y}\in S} \|\mathbf{F}(\mathbf{y}) - \mathbf{T}^i_j\|$
    if $\|\mathbf{F}(\mathbf{y}^*_j) - \mathbf{T}^i_j\| > 0$ **then**
      $\mathbf{y}_{|P|+1} := \mathbf{y}^*_j$
      $P := P \cup \{\mathbf{y}_{|P|+1}\}$
      $FP := FP \cup \{\mathbf{F}(\mathbf{y}_{|P|+1})\}$
    end if
  end for
end for

Here, $M \in \mathbb{N}$ is a predefined maximum number of steps, P is the resulting set that approximates $\mathscr{P}$ (at least in parts) and FP is the corresponding front. The *if*-condition ensures that the chosen targets are not reachable, at least for the chosen local minimizer. For target generation, we have chosen a strategy designed to approximate the Pareto front for a bi-criteria optimization problem with evenly spread points very quickly. Details can be found in [133]. Different possibilities relating to the generation of good targets are proposed in e.g. [33].

3.2.4.2 Description

A general formulation of a multiobjective optimal control problem with uncertainty is stated as follows.

Problem 3.2. Find a control $\mathbf{u}^*$ that solves

$$\min_{\mathbf{u}} \hat{\mathbf{J}}(\mathbf{u}) = \min_{\mathbf{u}} \int \mathbf{J}(\mathbf{x},\mathbf{u})\rho(\xi)\,d\xi \tag{3.11}$$

subject to

$$\dot{\mathbf{x}}(t) = \mathbf{f}(\mathbf{x}(t),\mathbf{u}(t),\xi), \tag{3.12}$$

with vector-valued objective functionals $\hat{\mathbf{J}} = (\hat{J}_1,\dots,\hat{J}_k)$ and $\mathbf{J} = (J_1,\dots,J_k)$ such that $\hat{J}_i(\mathbf{u}) = \int J_i(\mathbf{x},\mathbf{u})\rho(\xi)\,d\xi$. Each $J_i(\mathbf{x},\mathbf{u})$ may consist of a Lagrange and a Mayer term of the form $J_i(\mathbf{x},\mathbf{u}) = \int_I C_i(\mathbf{x}(t),\mathbf{u}(t))\,dt + \psi_i(\mathbf{x}(t_f))$, $i = 1,\dots,k$, with $\mathbf{C} = (C_1,\dots,C_k)$ and $\psi = (\psi_1,\dots,\psi_k)$ being vector-valued, continuously differentiable functions.

The numerical solution to Problem 3.2 involves three key tasks: (i) the numerical solution of the state equation (3.6), (ii) a simulation strategy to evaluate the objective functional $\hat{\mathbf{J}}$, and (iii) an optimization algorithm to approximate the Pareto set.

For the first task, a numerical integrator based on a prescribed time grid $\Delta t = \{t_0,\dots,t_N = t_f\}$ is used. Let $\mathbf{u}_d = \{\mathbf{u}_k\}_{k=0}^N$ be a discretization of the time-dependent function $\mathbf{u}(t)$, where $\mathbf{u}_k$ is an approximation of $\mathbf{u}(t_k)$. For a fixed control sequence $\mathbf{u}_d$ and fixed parameter ξ a variational integrator is applied to compute an approximation $\mathbf{x}_d = \{\mathbf{x}_k\}_{k=0}^N$ of the curve $\mathbf{x}(t)$.[9] Variational integrators [97] are particularly

[9] Note that each x_k, $k = 0,\dots,N$, is a function of ξ.

efficient for Hamiltonian and Lagrangian systems, since they preserve structural properties such as symmetries in the discrete approximation.

For the second task, in general two approximations have to be performed in order to evaluate an objective functional of the form (3.11). In the following, we use the same approximation rules for each objective functional $\hat{J}_i$, $i = 1,\dots,k$ but omit the index i and the Mayer term for simplicity in this section. Based on the discrete time grid $\{t_k\}_{k=0}^N$, first the objective functional J is approximated by

$$J_d(\mathbf{x}_d(\xi),u_d) = \sum_{k=0}^{N-1} C_d(\mathbf{x}_k,\mathbf{x}_{k+1},\mathbf{u}_k) \approx \int_I C(\mathbf{x}(t,\xi),\mathbf{u}(t))\,dt \tag{3.13}$$

where $C_d(\mathbf{x}_k,\mathbf{x}_{k+1},\mathbf{u}_k)$ is an approximation of $\int_{t_k}^{t_{k+1}} C(\mathbf{x}(t,\xi),\mathbf{u}(t))\,dt$ for a fixed parameter $\xi = (\xi_1,\dots,\xi_s) \in \mathbb{R}^s$ using numerical quadrature rules (see e.g. [107]). Thus, $\int J_d(x_d(\xi),u_d)\rho(\xi)\,d\xi$ is already an approximation of $\hat{J}$ based on the discrete time grid $\{t_k\}_{k=0}^N$.

For the approximation of the remaining integral, we use a straightforward approximation technique based on a simple sampling of the uncertain parameters. We assume a bounded support of the pdf which is effective by cutting the pdf at appropriate boundaries. For our numerical examples, such a support is chosen to be five times the standard deviation, since it is unlikely that the parameters are outside this region. Note that also, in most practical applications, the region of uncertainty is bounded (system parameters can only change within a prescribed region). Thus, we introduce lower and upper bounds ξ_i^l and ξ_i^u on the random variables, such that $\xi_i \in [\xi_i^l,\xi_i^u]$ for $i = 1,\dots,s$. We discretize ξ with $M_i + 1$ discretization points for each component ξ_i, i=1,...,s. Thus, we define equidistant discrete grids $\{\xi_{ik_i}\}_{k_i=0}^{M_i}$ with $\xi_{i0} = \xi_i^l$, $\xi_{ik_i} = \xi_{i0} + \Delta\xi_i k_i$ and $\xi_{iM_i} = \xi_{i0} + M_i\Delta\xi_i = \xi_i^u$ with grid sizes $\Delta\xi_i \in \mathbb{R}$, $i = 1,\dots,s$.[10] We define a discretized probability density function ρ_d as

$$\rho_d(\xi) = \frac{\rho((\xi_{1k_1},\cdots,\xi_{sk_s}))}{\Delta\xi_1\cdots\Delta\xi_s \sum_{l_1=1}^{M_1}\cdots\sum_{l_s=1}^{M_s}\rho((\xi_{1l_1},\cdots,\xi_{sl_s}))} \tag{3.14}$$

$$\text{for}\quad \xi_{ik_i-1} < \xi_i \le \xi_{ik_i} \quad \text{for} \quad i = 1,\dots,s,$$

$k_i = 1,\dots,M_i$, $i = 1,\dots,s$ and $\rho_d(\xi) = 0$ if $\xi_i \notin [\xi_i^l,\xi_i^u]$ for at least one i. Note that this choice of ρ_d guarantees that $\int \rho_d(\xi)\,d\xi = 1$. Finally, $\hat{J}(u)$ is approximated by

$$\begin{aligned}\hat{J}_d(u_d) &= \Delta\xi_1\cdots\Delta\xi_s \sum_{l_1=1}^{M_1}\cdots\sum_{l_s=1}^{M_s} J_d(x_d(\xi_{1l_1},\cdots,\xi_{sl_s}),u_d)\rho_d((\xi_{1l_1},\cdots,\xi_{sl_s}))\\ &= \frac{\sum_{l_1=1}^{M_1}\cdots\sum_{l_s=1}^{M_s} J_d(x_d(\xi_{1l_1},\cdots,\xi_{sl_s}),u_d)\rho((\xi_{1l_1},\cdots,\xi_{sl_s}))}{\sum_{l_1=1}^{M_1}\cdots\sum_{l_s=1}^{M_s}\rho((\xi_{1l_1},\cdots,\xi_{sl_s}))}. \end{aligned} \tag{3.15}$$

[10] In the following, the first index of ξ refers to the vector component, while the second index represents the grid point.

Note that the probability P of ξ being in $[\xi_{1k_1-1}, \xi_{1k_1}] \times \cdots \times [\xi_{sk_s-1}, \xi_{sk_s}]$ can be approximated by

$$P_{k_1 \cdots k_s} = \int_{\xi_{1k_1-1}}^{\xi_{1k_1}} \cdots \int_{\xi_{sk_s-1}}^{\xi_{sk_s}} \rho_d(\xi)\, d\xi = \frac{\rho((\xi_{1k_1}, \cdots, \xi_{sk_s}))}{\sum_{l_1=1}^{M_1} \cdots \sum_{l_s=1}^{M_s} \rho((\xi_{1l_1}, \cdots, \xi_{sl_s}))}. \tag{3.16}$$

The multiobjective optimal control problem is thus transformed into a multiobjective optimization problem with objective function $\hat{\mathbf{J}}_d = (\hat{J}_{d,1}, \ldots, \hat{J}_{d,k})$ and optimization parameters $\mathbf{y} = \mathbf{u}_d = \{\mathbf{u}_0, \ldots, \mathbf{u}_N\}$. The discretization of the time interval determines the problem dimension and is typically high in order to comply with the desired accuracy requirements. Thus, for the computation of Pareto optimal control sequences $\mathbf{u}_d = \{\mathbf{u}_0, \ldots, \mathbf{u}_N\}$ and for the approximation of the Pareto set (the third task), the reference point optimization method is applied. The discretization of the uncertain parameter space significantly influences the duration of the objective function evaluation. Note that for each evaluation of $\hat{\mathbf{J}}_d(\mathbf{u}_d)$, $(M_1 + 1) \cdots (M_s + 1)$ simulations of the differential equation (3.12) are required.

3.2.4.3 Results

As a result, a set of Pareto optimal control sequences $\mathbf{u}_d$ is obtained. By applying these control sequences to the technical system described by the differential equation (3.6), the system behaves Pareto optimally with respect to the prescribed quantity of interest (e.g. control effort) and performance in the presence of uncertain parameters.

3.2.4.4 Application Example

In this section, a robot arm modeled as a two-link mechanism (see Fig. 3.23) is considered for which the lengths L_1 and L_2 of the two links are assumed to be unknown or not measurable exactly. The mechanism consists of two coupled planar rigid bodies, where θ_i, $i = 1, 2$, denote the orientation of the ith link measured counterclockwise from the positive horizontal axis. The system is controlled via two control torques denoted with $\mathbf{u}(t) = (\tau_1(t), \tau_2(t))$, acting in both joints of the two time-dependent links (see [107] for a detailed model description). The goal is to determine a control sequence $\mathbf{u}(t)$ which steers the robot arm tip (modeled as the end point of the two-link mechanism) from a prescribed initial state $\mathbf{x}_0$ to a

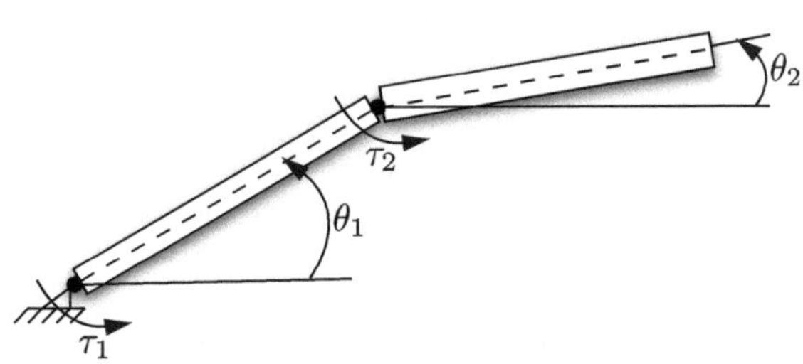

Fig. 3.23 Model of the two-link manipulator with uncertain link lengths

prescribed final state $\mathbf{x}_{\text{ref}} = (0, 1.5, 0, 0)$ at time $t_f = 1.0$, where $\mathbf{x}$ consists of the arm tip's position $\mathbf{q} = (L_1 \cos\theta_1 + L_2 \cos\theta_2, L_1 \sin\theta_1 + L_2 \sin\theta_2)^T$ and its velocity $\mathbf{v} = (-L_1\dot{\theta}_1 \sin\theta_1 - L_2\dot{\theta}_2 \sin\theta_2, L_1\dot{\theta}_1 \cos\theta_1 + L_2\dot{\theta}_2 \cos\theta_2)^T$. This maneuver should be performed in such a way that, on one hand, the required control effort is minimized, and on the other hand, the deviation from the prescribed goal state $\mathbf{x}_{\text{ref}}$ is as small as possible in the presence of uncertain length parameters $\xi = (L_1, L_2)$.[11]

In the first step, the deterministic optimal control problem is solved, for which the lengths are fixed as some reference value $\xi = \hat{\xi} = (\hat{L}_1, \hat{L}_2)$. The endpoint condition $\mathbf{x}(t_f) - \mathbf{x}_{\text{ref}} = 0$ is incorporated as a terminal constraint in the optimal control problem, such that only one objective functional, the control effort given as $J_1(u) = \int_{t_0}^{t_f} \frac{1}{2}\|\mathbf{u}(t)\|_2^2 dt$, is considered. To determine the optimal control sequence $\mathbf{u}_d^*(\hat{\xi})$, the numerical method DMOC (Discrete Mechanics and Optimal Control) [107] is utilized, which is based on the discretization of the underlying optimal control problem. For the discretization, a discrete time grid with $N+1 = 20$ discretization points is chosen.

In a second step, the control problem is reconsidered including uncertainty. The unknown lenghts L_1 and L_2 are assumed to be independent and normally distributed around the reference value $\hat{L}_i = 1.0$, $i = 1, 2$, with a standard deviation of 0.001. For the numerical treatment, the space of uncertain parameters $[0.995, 1.005] \times [0.995, 1.005]$ is discretized by two equidistant grids $\{\xi_{i0}, \xi_{i1}, \dots, \xi_{iM_i}\}$, $M_i = 33$, $i = 1, 2$, and we approximate the pdf with the discretized probability density function $\rho_d(\xi)$ (3.14). In Fig. 3.24 (left), the probability P of ξ being in $[\xi_{1k-1}, \xi_{1k}] \times [\xi_{2l-1}, \xi_{2l}]$ is depicted as given by P_{kl} in (3.16).

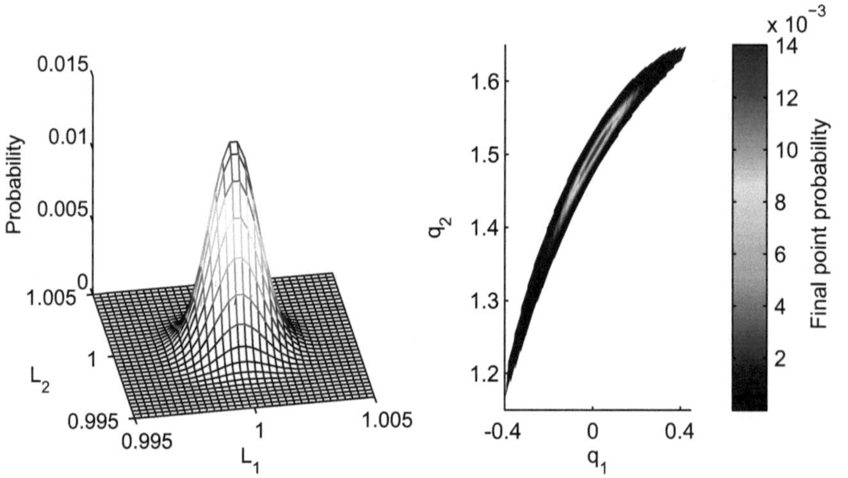

Fig. 3.24 Left: discretization of uncertain parameter space and approximated probability; Right: final positions $q(t_f, \xi)$ for different values of lengths parameters ξ and optimal control sequence $\mathbf{u}_d^*(\hat{\xi})$

[11] Note that the robot arm tip's state is dependent on the uncertain lengths ξ.

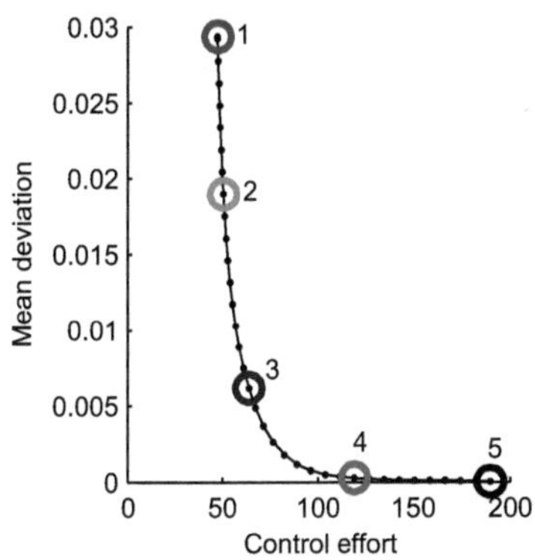

(a) Approximation of Pareto front with five chosen points.

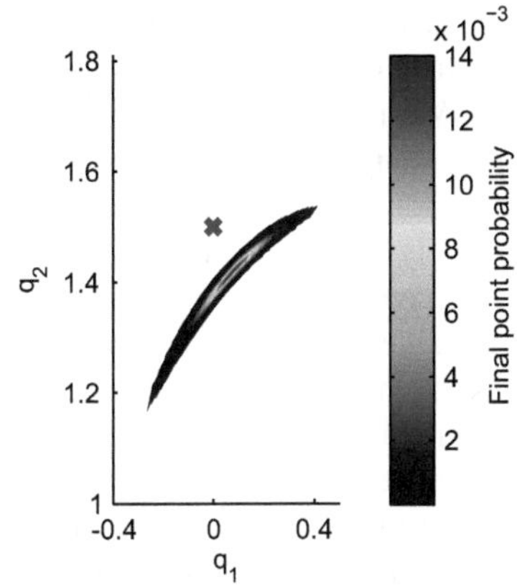

(b) Final configurations for Pareto point 1.

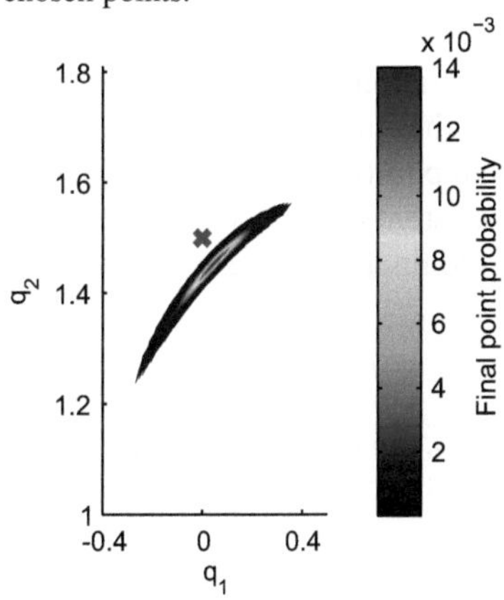

(c) Final configurations for Pareto point 2.

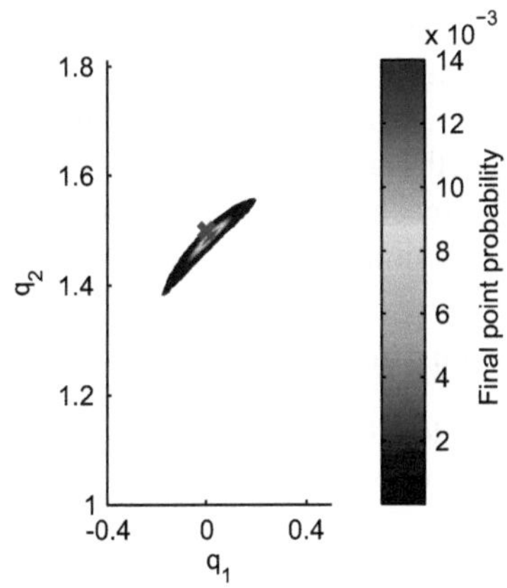

(d) Final configurations for Pareto point 3.

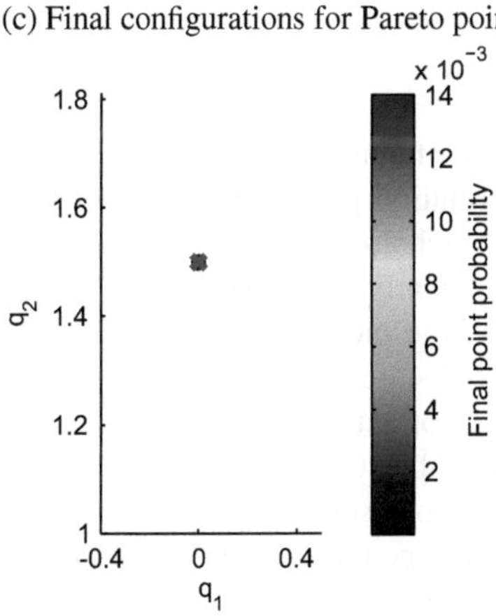

(e) Final configurations for Pareto point 4.

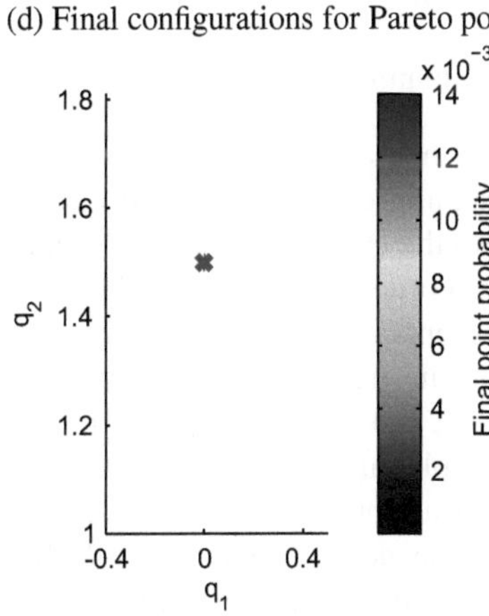

(f) Final configurations for Pareto point 5.

Fig. 3.25 Behavior of final points in configuration space for five chosen points of the approximated Pareto front

First, we investigate the influence of uncertainty on the system performance if the optimal control $\mathbf{u}_d^*(\hat{\xi})$, which was computed for the deterministic case, is applied to the system. We integrate the differential equation (3.6) with fixed optimal control sequence $\mathbf{u}_d = \mathbf{u}_d^*(\hat{\xi})$ and different grid values $\xi_{ij} = (\xi_{1i}, \xi_{2j})$, $i, j = 0, \ldots, M$. In Fig. 3.24 (right), it can be seen that the different final configurations $\mathbf{q}(t_f, \xi_{ij})$ of the arm tip strongly deviate from the reference value for the final position $\mathbf{q}_{\text{ref}} = (0, 1.5)$. A similar behavior can be observed for the final velocities. The colors indicate the probability values as given in Fig. 3.24 (left). Thus, the performance of the system is relatively poor in the presence of uncertain lengths and new control sequences have to be determined to improve the system performance.

To reduce the deviation from the reference goal state, the mean of the performance measure (3.8) is treated as additional objective function, and the reference point optimization method is employed to determine Pareto optimal control sequences (minimal control effort and minimal mean deviation from the reference point). For the minimization of (3.10), a SQP method implemented in NAG is used.

Figure 3.25 (a) shows the approximation of the Pareto front. It can be observed that for increasing control effort, the mean of the deviation from the final goal state (marked as a cross) decreases. In Fig. 3.25 (b)–(f), the final configurations for uncertain lengths are illustrated for five selected Pareto points. Along the Pareto front, the area of reached final points contracts and thus system performance increases as desired.[12] A similar behavior can be observed for the final velocities.

3.2.4.5 Further Reading

For a more detailed description of numerical methods, the problem formulation and an overview of related literature, we refer to [107, 108, 135].

3.2.5 Behavior Planning

Philip Hartmann

In order to increase the dependability of self-optimizing mechatronic systems, cognitive planning components with enhanced information processing are also integrated into the system. These components allow mechatronic systems to plan their behavior in order to fulfill individual tasks independently and proactively. A single task represents a sequence of actions executed by the mechatronic system within a limited time frame in order to reach a given goal state. Along with bare fulfillment of that task, i.e. finding an arbitrary sequence of actions to reach the desired goal-state, planning tries to minimize or maximize objectives, such as minimizing energy consumption. For this reason, actions are only selected if their expected results fit the desired objectives. With respect to dependability, it is possible to create alternative plans for critical situations before they arise, i.e. for particular

[12] For the illustration of the distributions in Fig. 3.24 (right) and Fig. 3.25 (b)-(f), only the final points with probability $P \geq 10^{-6}$ are depicted.

environmental or low energy situations. However, this may decrease the Availability of the mechatronic system and the Reliability of subsequent task fulfillment. Furthermore, behavior planning considers the continuous and nondeterministic environment of the system (cf. [80]).

3.2.5.1 Prerequisites and Input

When modeling a planning domain for behavior planning of intelligent mechatronic systems (cf. [80–82]), the main challenge is to map the partial function solutions onto actions within the framework of PDDL (Planning Domain Definition Language, cf. [53]). Depending on the amount of detail desired when modeling these functions, this approach results in a higher or lower abstraction of actions. In case the of behavior planning, the executed partial function solutions are called operation modes. Thus, a planning problem for mechatronic systems can be formulated as follows (adapted from [81]):

- OM is a finite set of available operation modes,
- S is a finite set of possible system states, and
- $\mathbf{s} \in S$ is a state vector with $s(i) \in \mathbb{R}$ for the i-th component.

Furthermore, for each operation mode $om \in OM$:

- $prec^{om} := \{(x_{lower} < s(i) < x_{upper}) | x_{lower}, x_{upper} \in \mathbb{R}\}$ is the set of preconditions which must be true for the execution of operation mode om and
- $post^{om}$ is a set of conditional numerical functions describing the change of influenced state variables. A condition is a logical expression (conjunctions and disjuctions) of comparison operations; if a condition is true, the result of the corresponding numerical function is assigned to state variable in the next state $\mathbf{s}'$ of the plan ([81]).

3.2.5.2 Description

A specific planning problem is the finding of a sequence of operation modes which describes a transition from an initial system state $\mathbf{s}_i \in S$ to a predetermined goal state $\mathbf{s}_g \in S$. Thus, a single task of a mechatronic system is given as a 2-tuple $O = (\mathbf{s}_i, \mathbf{s}_g)$. A solution to the planning problem can be determined by applying a state space search algorithm (cf. [57]), for example. The optimal solution (e.g. minimum of energy consumption) can be found by computing the specific solutions with respect to the given System of Objectives. For this purpose, Ω is a set of objectives and $f : S \times \Omega \to [0, 1]$ is a function that indicates how well the execution of an operation mode in a given state satisfies the objective. Using the weighted sum of the objectives, the optimal sequence of operation modes can be determined (cf. [81]).

During runtime in a non-deterministic environment with continuous processes, behavior planning has to include methods for handling resulting problems. For example, Klöpper [80] uses a modeling approach to integrate continuous processes based on optimal control and continuous multiobjective optimization (also cf. [56]), as well as estimation obtained by fuzzy approximation. To manage planning under uncertain conditions, different techniques can be combined in a hybrid planning architecture (cf. [81]).

Figure 3.26 shows the hybrid planning architecture with the corresponding components for planning, execution and monitoring of plans. The total planning is divided into three separate sections: offline, just-in-case and online planning. The offline planning represents a planning process where, initially, a deterministic and optimal plan in view of the objectives is fully created before execution. The resulting plan is used in the just-in-case planning to do a probabilistic analysis for plan deviations. The present and deterministic plan is examined for estimated variances in order to proactively generate conditional branches, with alternative plans for critical system conditions. A threshold specifies the maximum probability of state deviations which would results in a generation of conditional branches (See [82], cf. in particular also [82] and [80].)

For this purpose, an additional stochastic planning model is formulated based on the deterministic planning model. This consists of stochastic states $\mathbf{s}^p$ with $|\mathbf{s}^p| = |\mathbf{s}|$, where $range(s^p(i)) \to P(\mathbb{R})$ is the values range and $distribution(s^p(i))$ the probability distribution of the state variable $s^p(i)$ and a stochastic variant of the operation modes. Let $in_s^{om} \subseteq pre^{om}$ be a subset of input variables and $out_s^{om} \subseteq post^{om}$ a subset of output variables. For each output variable $o \in out_s^{om}$, a Bayesian network (cf. [17]) bn_o^{om} is created to formulate the stochastic relation (cf. [80, 81]; for a concrete example of creating a stochastic model cf. [82]). As a result, it is now possible to use the just-in-case-planning to generate alternative plans for situations that could occur with high probability during operation.

The online planning (cf. Fig. 3.26) serves as a fallback mechanism; it selects the optimal operation mode for the next execution step. Thus, operation in previously unplanned situations is guaranteed. A simulation of the continuous system behavior will check whether the current action of the active plan is executable under the given environmental conditions. If this is not possible, online planning is necessary, e.g. for a situation with extreme environmental influences such as heavy rain. While

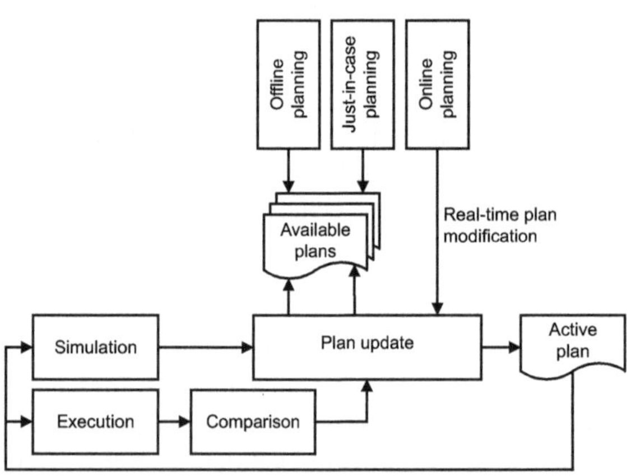

Fig. 3.26 Hybrid planning architecture (source: [81])

completing the execution of previously planned operation modes, a comparison of planned and actually reached system states is carried out.

A process for plan updating will check whether a pre-determined plan is available or whether a plan modification by the online planning is necessary. This will guarantee the immediate availability of the next operation mode (cf. Fig. 3.26).

The just-in-case and online planning are implemented as anytime algorithms (for the usage of anytime algorithms in intelligent systems cf. [155]). The planning process can be interrupted at any time to obtain a result, but with increasing time for calculations it provides a higher quality of result, as it is possible to generate more branches and to reach a higher depth of planning.

3.2.5.3 Results

The dependability our type of system can be influenced by various factors. A major factor is the availability of energy, as this is crucial for the operationf of the system. To ensure the dependability of the mechatronic system, it is essential to use the energy storage in a valid range and in particular to observe the state of charge continuously. Energy management can use behavior planning to proactively schedule future energy demands according to the fulfillment of the current task, which increase the dependability of the mechatronic system (cf. [82]).

Table 3.2 Values for f_1 (weighted average body acceleration in m/s^2) and f_2 (energy consumption in ws) of operation modes derived from the multiobjective optimization of the active suspension module. (source: [81])

OM	Objective function	Track type									
		I	II	III	IV	V	VI	VII	VIII	IX	X
a	f_1	0.117	0.233	0.350	0.466	0.583	0.699	0.816	0.932	1.049	1.166
	f_2	196	393	589	786	982	1179	1375	1572	1768	1965
b	f_1	0.152	0.304	0.457	0.609	0.761	0.913	1.066	1.218	1.370	1.522
	f_2	165	329	494	659	823	988	1153	1317	1482	1647
c	f_1	0.192	0.385	0.577	0.770	0.962	1.155	1.347	1.540	1.732	1.925
	f_2	142	283	425	567	709	850	992	1134	1275	1417
d	f_1	0.224	0.449	0.673	0.897	1.122	1.346	1.570	1.794	2.019	2.243
	f_2	122	245	367	489	612	734	856	979	1101	1224
e	f_1	0.262	0.523	0.785	1.047	1.308	1.570	1.832	2.093	2.355	2.617
	f_2	104	208	313	417	521	625	730	834	938	1042
f	f_1	0.298	0.595	0.893	1.191	1.488	1.786	2.084	2.381	2.679	2.977
	f_2	87	173	260	346	433	520	606	693	779	866
g	f_1	0.331	0.662	0.994	1.325	1.656	1.987	2.318	2.649	2.981	3.312
	f_2	69	138	206	275	344	413	482	550	619	688
h	f_1	0.375	0.749	1.124	1.499	1.873	2.248	2.623	2.997	3.372	3.747
	f_2	50	99	149	199	248	298	348	398	447	497
i	f_1	0.435	0.870	1.305	1.739	2.174	2.609	3.044	3.479	3.914	4.349
	f_2	27	55	82	110	137	164	192	219	247	274

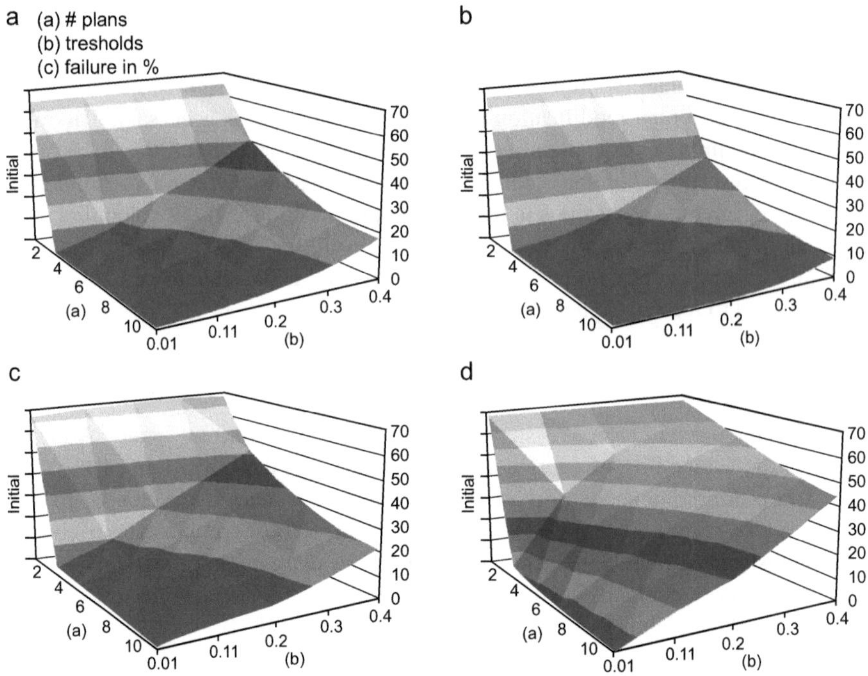

Fig. 3.27 Percentage of failed execution depending on threshold probability and number of available alternative plans: (a) Return to standard plan (±0%); (b) No return to standard plan (±0%); (c) Return to standard plan (±15%); (d) Return to standard plan (+15%) (source: [81])

3.2.5.4 Application Example

Table 3.2 shows the values for operation modes derived from the multiobjective optimization from the active suspension module of the RailCab system.

The experiments described here were intended to allow an evaluation of three hypotheses (cf. [81]). One of these hypotheses in connection with the dependability was that a lower threshold probability and a higher number of alternative plans increases the reliability of the just-in-case planning ([82]). The simulated experiments included four scenarios (source [81]):

1. (±0%): The energy consumptions drawn from track networks were not changed during simulation.
2. (±15%): The energy consumptions drawn from track networks were either decreased or increased by a random value up to 15%.
3. (+15%): The energy consumptions drawn from track networks were always decreased by a random value up to 15%.
4. (−15%): The energy consumptions drawn from track networks were always increased by a random value up to 15%.

The Results are shown in Fig. 3.27 (for a detailed description of the simulation parameters and the executed scenarious cf. [81]) When regarding the percentage of failed plan execution during the simulation runs for different scenarios, adjusting the two parameters threshold value and number of alternative plans reduces the number of failed plans significantly.

3.2.5.5 Further Reading

A detailed explanation of behavior planning for mechatronic systems can be found in [81, 82]. In particular, [82] gives a deeper understanding of the probabilistic plan structure used in the analysis of the just-in-case planning. The basic methods were originally published in the dissertation [80], which may also be a good starting point for further reading.

3.2.6 Computation of Robust Pareto Points

Michael Dellnitz, Robert Timmermann, and Katrin Witting

During the development of self-optimizing systems, the technical system under consideration usually has to be optimized with respect to several different objectives. Typically, these objectives are in conflict with each other, such as safety and energy efficiency, for instance. In mathematical terms, the problems to be solved in this case are *multiobjective optimization* problems. Here, the attributes of dependability, in particular safety and reliability, can be considered as objective functions. The solution to multiobjective optimization problems is given by the set of optimal compromises, the previously introduced *Pareto set*. The elements of this set define the respective status of the system and are called *Pareto points*. A multiobjective optimization method is often applied in combination with control engineering methods. It requires the same information about the models of the dynamics from the appropriate partial models of the Principle Solution (cf. Sect. 2.1). Of particular importance is the partial model "System of Objectives". It contains information about the relevant objectives which have to be considered to solve the multiobjective optimization problem.

Both during the system's design phase and during operation, it is an important concern to choose a Pareto point which is appropriate for the current system environment. For instance, the objective "safety" should receive a higher priority when a moving vehicle is operating in a rainy or windy environment. Thus, the partial model "Environment" provides important information on environmental parameters and their ranges which has to be considered in the multiobjective optimization methods. In general, the choice of Pareto points must be adapted to the variation of external parameters.

In this section, parametric multiobjective optimization problems are considered in which an external parameter influences the system's behavior; this parameter may vary during runtime. We introduce here two methods which allow the computation of so-called *robust Pareto points*. These points are characterized by minimal variation with respect to changes of external parameters. The first method is based on the calculus of variations; the goal of this method is to identify a Pareto point which changes very little when the external parameter is varied over an entire interval. The second method is based on numerical path-following techniques. Here, a local strategy is used in order to update the status of the system in response to the variation of an external parameter.

3.2.6.1 Prerequisites and Input

Multiobjective Optimization

Consider an unconstrained multiobjective optimization problem (cf. also Sect. 3.2.4.1) which additionally depends on an external parameter $\lambda \in \mathbb{R}$. This parameter is not intended for optimization, but it nonetheless influences the system objectives. This unconstrained parametric multiobjective optimization problem can be formulated as

$$\min_{\mathbf{y}}\{\mathbf{F}(\mathbf{y},\lambda) : \mathbf{y} \in \mathbb{R}^n, \lambda \in [\lambda_{start}, \lambda_{end}]\}$$

where $\mathbf{F}$ is defined as the vector of objective functions $f_1,\ldots,f_k, k > 1$,

$$\mathbf{F} : \mathbb{R}^n \times \mathbb{R} \to \mathbb{R}^k, \mathbf{F}(\mathbf{y},\lambda) = (f_1(\mathbf{y},\lambda),\ldots,f_k(\mathbf{y},\lambda))^T .$$

A point $\mathbf{y}^* \in \mathbb{R}^n$ is called *Pareto optimal* for a given parameter λ, if there exists no $\mathbf{y} \in \mathbb{R}^n$ with

$$\mathbf{F}(\mathbf{y},\lambda) \leq_p \mathbf{F}(\mathbf{y}^*,\lambda) \text{ and } f_j(\mathbf{y},\lambda) < f_j(\mathbf{y}^*,\lambda) \text{ for at least one } j \in \{1,\ldots,k\}.$$

The set of all Pareto points is the *Pareto set*. A necessary condition for Pareto optimality is given by the (in this case parameter-dependent) *Kuhn-Tucker equations* (cf. [87]): for each Pareto optimal point $\mathbf{y} \in \mathbb{R}^n$ there exists a vector $\alpha(\lambda) = (\alpha_1(\lambda),\ldots,\alpha_k(\lambda))^T \in \mathbb{R}^k$ with $\alpha_i(\lambda) > 0$ such that

$$\mathbf{H}_{KT}(\mathbf{y}(\lambda),\alpha(\lambda),\lambda) = \begin{pmatrix} \sum_{i=1}^{k} \alpha_i(\lambda)\nabla_{\mathbf{y}} f_i(\mathbf{y}(\lambda),\lambda) \\ \sum_{i=1}^{k} \alpha_i(\lambda) - 1 \end{pmatrix} = 0. \quad (3.17)$$

The set of all $\mathbf{y}$ for which there exists a weight vector α, such that $(\mathbf{y},\alpha,\lambda)$ is a solution of (3.17), is called the *set of substationary points* S_λ. In numerical computations, it is often easier to work with $\alpha_i = t_i^2$ and solve $\mathbf{H}(\mathbf{y},\mathbf{t},\lambda)$, but both approaches compute the same set S_λ and thus t_i and α_i will be used synonymously throughout this section.

3.2.6.2 Description

In this section, the two methods (relying on variational calculus or numerical path-following techniques) used here for the computation of robust Pareto points are introduced in more detail.

Variational Approach

Our goal is to determine a curve $\gamma(\lambda) = (\mathbf{y}(\lambda), \mathbf{t}(\lambda))^T$ of minimal length from an arbitrary starting point on the set of substationary points $S_{\lambda_{start}}$ to an arbitrary end point on $S_{\lambda_{end}}$ that lies within the λ-dependent set of substationary points. Using calculus of variations, this problem can be formulated as follows:

$$\min_{\gamma} \int_{\lambda_{start}}^{\lambda_{end}} \|\mathbf{y}'(\lambda)\|_2 \, d\lambda \text{ s.t. } \mathbf{H}_{KT}(\mathbf{y}(\lambda), \mathbf{t}(\lambda), \lambda) = 0. \tag{3.18}$$

This functional calculates the length of the curve $\mathbf{y}(\lambda)$, which is guaranteed to lie on the set of substationary points.

A necessary condition for the optimality of (3.18) is given by the *Euler-Lagrange equations* (cf. [58]). In [151, 152] a discrete formulation of the Euler-Lagrange equations is described which goes back to [97]. This leads to a system of nonlinear equations that characterize candidates for robust Pareto points:

$$\begin{aligned}
\mathbf{H}_{KT}(\mathbf{y}_j, \mathbf{t}_j, \lambda_j) &= 0 \,\forall\, j = 0, \ldots, N \\
\mu_j^T \frac{\partial}{\partial t_j} \mathbf{H}_{KT}(\mathbf{y}_j, \mathbf{t}_j, \lambda_j) &= 0 \,\forall\, j = 0, \ldots, N \\
\frac{\mathbf{y}_{j+1} - 2\mathbf{y}_j + \mathbf{y}_{j-1}}{h^2} \mu_j^T \frac{\partial}{\partial \mathbf{t}_j} \mathbf{H}_{KT}(\mathbf{y}_j, \mathbf{t}_j, \lambda_j) &= 0 \,\forall\, j = 1, \ldots, N-1 \\
\frac{\mathbf{y}_1 - \mathbf{y}_0}{h^2} - \frac{1}{2} \mu_0^T \frac{\partial}{\partial \mathbf{t}_0} \mathbf{H}_{KT}(\mathbf{y}_0, \mathbf{t}_0, \lambda_{start}) &= 0 \\
-\frac{\mathbf{y}_N - \mathbf{y}_{N-1}}{h^2} - \frac{1}{2} \mu_N^T \frac{\partial}{\partial \mathbf{t}_N} \mathbf{H}_{KT}(\mathbf{y}_N, \mathbf{t}_N, \lambda_{end}) &= 0
\end{aligned} \tag{3.19}$$

where $N+1$ is the number of discretization points on the curves $\mathbf{y}(\lambda)$, $\mathbf{t}(\lambda)$ and $\mu(\lambda)$. μ is the Lagrangian multiplier. This system of equations (3.19) can be solved using numerical techniques in order to compute robust Pareto points. An application of this method for the RailCab's active suspension system can be found in Sect. 3.2.6.4.

Numerical Path-Following Approach

Given an initial Pareto set, we investigate if there exist points $\mathbf{y} \in \mathbb{R}^n$ that are substationary points for all $\lambda \in [\lambda_{start}, \lambda_{end}]$; if this is not the case we investigate which points $\mathbf{y} \in \mathbb{R}^n$ do not vary significantly during variation of λ. Analogously, it can be examined which points vary as little as possible in the respective image space.

Numerical path-following techniques are used to track a Pareto point during variation of the parameter λ. A predictor-corrector method was adapted to follow a Pareto point on the set of substationary points S (cf. [3] or [37] for an introduction to numerical path-following and the predictor-corrector method).

For this task, we are searching for the set of zeros $c(s)$ of a continuous function $\mathbf{H} : \mathbb{R}^n \times \mathbb{R} \to \mathbb{R}^n$. The aim of the predictor-corrector method is to calculate a sequence of points $(\mathbf{p}_i, \lambda_i)$, such that $\mathbf{H}(\mathbf{p}_i, \lambda_i) = 0$ for all $i = 1, 2, \ldots$. The numerical procedure consists of two steps, which are executed alternatingly: during the predictor step, a new point in the vicinity of the zero set is calculated; starting at this point, a new point on the zero set is calculated in the following corrector step.

Using necessary and sufficient conditions formulated by Luenberger in [96] and introducing Lagrangian multipliers μ, one can construct a path-following routine to compute paths containing substationary points for varying values of λ. These paths are computed in such a way that the distance from the previously computed point to the next point, which has to lie on the set of substationary points for the new λ-value, is minimal.

At first, a Pareto set for $\lambda = \lambda_{start}$ is computed (e.g. using the software GAIO, cf. [34]); for which $\mathbf{t}$ can be computed using the Kuhn-Tucker equations. We initially set $\mathbf{u} = (\mathbf{y}, \mathbf{t}, \lambda, \mu) = (\mathbf{y}_{start}, \mathbf{t}_{start}, \lambda_{start}, \mathbf{0})$.

Predictor Step

In this step λ is increased while $\mathbf{y}$, $\mathbf{t}$, and μ are left unchanged

$$\mathbf{u}_{Pred} = \mathbf{u} + \begin{pmatrix} \mathbf{0} \\ \mathbf{0} \\ h \\ \mathbf{0} \end{pmatrix} = \begin{pmatrix} \mathbf{y} \\ \mathbf{t} \\ \tilde{\lambda} \\ \mu \end{pmatrix}$$

where h is an adequately controlled stepsize.

Corrector Step

During the corrector step, $\tilde{\lambda}$ is fixed and $\mathbf{y}$, $\mathbf{t}$, and μ are adjusted until the point with minimal distance from the preceding set $S_{\tilde{\lambda}}$ is determined. The minimal distance computation can be formulated as a zero finding problem:

$$\mathbf{H}_{Corr}(\mathbf{y}, \mathbf{t}, \mu) = \begin{pmatrix} \sum_{i=1}^{k} t_i^2 \nabla_{\mathbf{y}} f_i(\mathbf{y}, \tilde{\lambda}) \\ \sum_{i=1}^{k} t_i^2 - 1 \\ \begin{pmatrix} \mathbf{d}(\mathbf{y}; \mathbf{y}_{old}, \lambda_{old}, \tilde{\lambda}) \\ \mathbf{0} \end{pmatrix} - \mu_1 \nabla_{(\mathbf{y},\mathbf{t})} \mathbf{H}_{KT}^1(\mathbf{y}, \mathbf{t}, \tilde{\lambda}) \ldots \\ \ldots - \mu_n \nabla_{(\mathbf{y},\mathbf{t})} \mathbf{H}_{KT}^n(\mathbf{y}, \mathbf{t}, \tilde{\lambda}) - \mu_{n+1} \begin{pmatrix} \mathbf{0} \\ 2t_1 \\ \vdots \\ 2t_k \end{pmatrix} \end{pmatrix} = 0$$

with $\mathbf{H}_{KT} = (\mathbf{H}_{KT}^1, \ldots, \mathbf{H}_{KT}^n)^T$. The function $\mathbf{d}(\cdot; \mathbf{y}_{old}, \lambda_{old}, \tilde{\lambda}) : \mathbb{R}^n \to \mathbb{R}^n$ is a decision function, which is defined as

(a) $\mathbf{x} \mapsto \mathbf{d}(\mathbf{x}; \mathbf{y}_{old}, \lambda_{old}, \tilde{\lambda}) = 2(\mathbf{x} - \mathbf{y}_{old})$ or

(b) $\mathbf{x} \mapsto \mathbf{d}(\mathbf{x}; \mathbf{y}_{old}, \lambda_{old}, \tilde{\lambda}) = 2\left(\sum_{i=1}^{k} \left(f_i(\mathbf{x}, \tilde{\lambda}) - f_i(\mathbf{y}_0, \lambda_{old})\right) \nabla_{\mathbf{x}} f_i(\mathbf{x}, \tilde{\lambda})\right)$.

The choice of version (a) or (b) depends on the space within which we are seeking paths of minimal length: if we consider paths of minimal length in pre-image space, version (a) is used, and if paths of minimal length in image space are considered, version (b) is used.

The predictor and the corrector step are repeated alternatingly until $\lambda = \lambda_{end}$, and thus a path from $S_{\lambda_{start}}$ to $S_{\lambda_{end}}$, has been found.

3.2.6.3 Results

Both methods, the variational approach as well as the path-following method, have been applied very successfully to realistic technical examples. The path-following approach has been used to compute robust Pareto points for the design of integrated circuits (ICs, cf. [23, 151]). In this context, the decision maker is interested in parameters for the IC design, which result in similar behavior of the IC across a wide range of external parameters (e.g. temperature or supply voltage); the concept of robust Pareto points can help identify these parameters. This approach has also been used to compute robust Pareto points for the RailCab's active suspension module in [145]. Another example of a technical application in which we use the variational approach to compute robust Pareto points is discussed in the following section.

3.2.6.4 Application Example

In [86], the variational method is used to compute robust Pareto points for the RailCab's active suspension module (ASM, cf. Sect. 1.3.1.3). An external parameter λ is used to model the varying crosswind conditions which affect the ASM. In this example, a simple sky-hook controller is used to control the system. The system depends on three parameters $\mathbf{p} = \{p_1, p_2, p_3\}$, which represent the damping of each degree of freedom of the coach body by the active suspension system. Furthermore, the two objectives *comfort* (f_1) and *energy consumption* (f_2) for the multiobjective optimization are defined as

$$f_{1,2} : \mathbb{R}^3 \to \mathbb{R}, \quad \mathbf{p} \mapsto f_{1,2}(\mathbf{p}) = \int_0^T \mathbf{y}(t)^T Q_{1,2} \mathbf{y}(t) dt \,,$$

with positive definite matrices $Q_{1,2}$. These objectives depend on the response $\mathbf{y}(t)$ of the linear model of the ASM, which itself depends on the parameters $\mathbf{p}$. The model is simulated with a fixed excitation $u(t)$ for a constant time T. Additionally, the crosswind is modeled as a further disturbance $z(t)$. An example crosswind profile is shown in Fig. 3.28.

We chose $N = 10$ discretization points from the mean crosswind λ between 0 m/s and 16.3 m/s. Based on this choice, the values of λ_j are given by $\lambda_j = 0 + j \cdot 0.163$, $j = 0, \ldots, N$. In this case, the system of equations (3.19) consists of 99 equations with 99 unknowns. Solutions were computed numerically using the MATLAB solver *fsolve* and are plotted as black points in Fig. 3.29.

The robust Pareto point at $(0,0,0)$ can be easily explained, as one would expect the energy-optimal solution at this point, regardless of the crosswind. The second

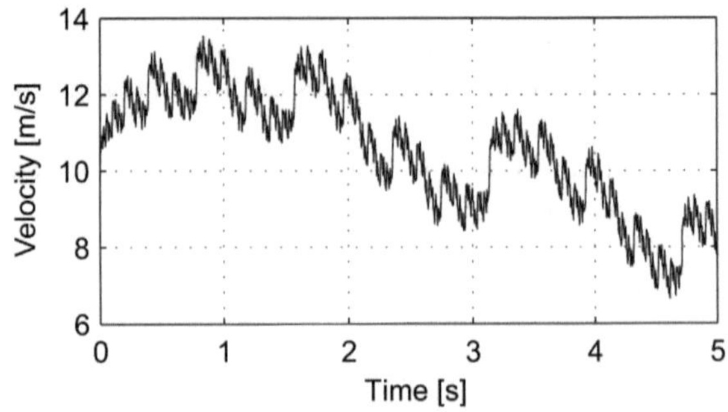

Fig. 3.28 One example: crosswind velocity profile acting as an external disturbance on the RailCab's active suspension module

robust Pareto point is nontrivial, though, and was not expected prior to the calculations. Thus, this provides some additional information about the active suspension system which might be used for the self-optimization process in future. The availability of robust Pareto points introduces a classification of optimal system configurations that can be used during system operation, for example when dependability is one of the design objectives.

3.2.6.5 Further Reading

A more detailed explanation of the variational approach can be found in [152], the path-following method has been published in [35], and the dissertation [151] gives a comprehensive overview of all topics covered in this section. The examples of the application of the numerical path-following method have been published

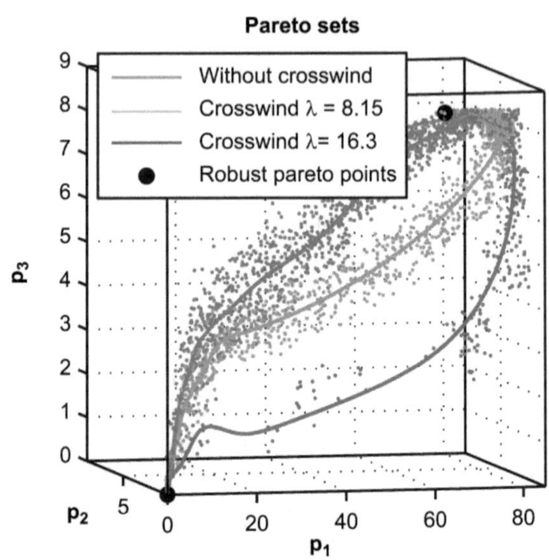

Fig. 3.29 Pareto sets for three specific crosswind values and robust Pareto points

in [23] (ICs) and [145] (RailCab) and the RailCab example of the variational method in [86].

3.2.7 Behavior-Based Adaptation of Differing Model Parameters

Bernd Kleinjohann, Lisa Kleinjohann, and Christoph Rasche

As discussed in previous chapters, continuous monitoring and behavior-based adaptation of self-optimizing mechatronic systems can considerably increase their dependability. Typically, such complex systems plan their actions as described in Sect. 3.2.5, for example. During the execution of a previously computed plan, the environment in which a mechatronic system is active may change in such a way that unsafe system states occur. Hence, in order to achieve dependability, environmental changes affecting a self-optimizing mechatronic system have to be taken into account to prevent system failures.

The method presented in this section has been designed to work with mechatronic systems that compute plans in order to move from one position to another. These plans consist of, for example, a set of Pareto points provided by multiobjective optimization [34, 135], as described in Sect. 3.2.6 and *D.M.f.I.T.S*, [55], Sect. 1.2.3. Similar to planning and multiobjective optimization the behavior-based adaptation of self-optimizing mechatronic systems is implemented in the Cognitive Operator of the OCM (cf. Sect. 1.1.1).

To ensure that a mechatronic system is able to move from an initial position to a destination, several conditions have to be considered. The passengers, for instance, most likely desire a certain level of comfort during operation, while only limited energy resources are available, and the destination has to be reached at some given time. These conditions then conflict with each other. To find optimal solutions, Pareto fronts can be calculated a priori, using a model of the system. Based on single points of the Pareto fronts, a plan can then be computed which ensures that all conditions are fulfilled as far as possible for the overall plan. Such a plan consists of several Pareto points, one for each part of the route. It is computed by a planner such as the one presented in *D.M.f.I.T.S*, [55], Sect. 5.3.8. Each Pareto point leads to specific parameter settings and influences how well the conditions, whether of primary importance, such as energy consumption, or of secondary importance, such as level of comfort, are fulfilled.

Computing Pareto fronts is of high computational complexity; therefore, the computation of the single Pareto fronts, used to determine specific settings for each track section, is performed prior to the first movement and stored in a database. This decreases the computational effort needed to compute a plan. The fronts in the database may be updated due to changing conditions.

As mentioned, the resulting plan is based on precalculated Pareto fronts. Hence, only the preliminary parameters of the mechatronic system model used to calculate the Pareto fronts could be taken into account while computing a workable plan. Therefore, deviations between the model or its parameters and the real system cannot be ruled out. One example is drag, which can lead to an increase or decrease

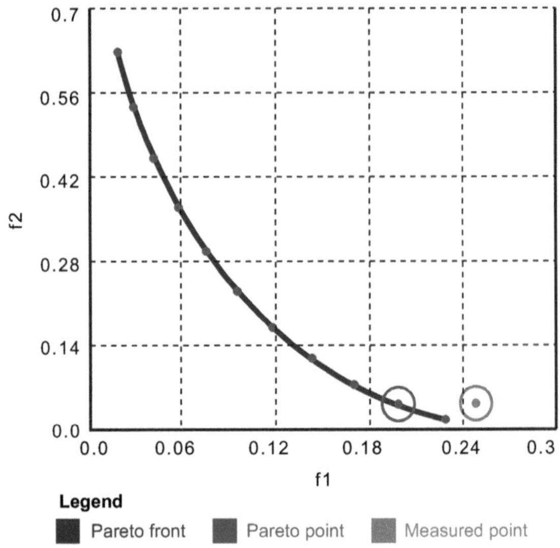

Fig. 3.30 The line shows a model Pareto front; the left point is the Pareto point selected by the planner. The right point is the determined working point, based on the current measurements.

in energy consumption of the real system compared to the energy consumption expected from the model values. To prevent the higher consumption from causing the mechatronic system to run out of energy before reaching its destination, a change in the plan has to be performed at runtime, thus possibly preventing a severe failure in system availability.

Figure 3.30 shows an example of such a case. The line denotes the Pareto front given by the model values and the point on the front denotes the Pareto point selected by the planner. Each Pareto point has a comfort value, given by the objective function $f1$, and an energy consumption value, given by $f2$. The momentary comfort and the energy consumption measured, result in the working point, i. e. the right point outlined by a circle. While there is only a small change in the energy consumption, the difference between the measured comfort value and the comfort value given by the Pareto front differ significantly. Such differences can invalidate the entire plan and a recalculation, based on the newly determined working point, has to be conducted.

To be able to detect such deviations and to change the plan, several values, such as energy consumption and passenger comfort, have to be measured continuously during operation. Based on the measured values, the current working point has to be calculated. If this working point differs too much from the model Pareto point which was selected by the planner, replanning is necessary to still be able to fulfill all conditions.

It is not possible to simply compute new Pareto fronts, which would lead to the measured working point. Each Pareto front is based on several parameters and if the model Pareto front does not fit the measurements, at least one of the parameters must have changed to an unknown value. This means that an approximation of the model Pareto front using parameter fitting to shift it closer to the measurements has to be conducted as described in the following sections.

3.2.7.1 Prerequisites and Input

The method *Behavior-Based Adaptation of Differing Model Parameters* needs several inputs. A plan and the corresponding Pareto points for the complete route are needed. Additionally, the multiobjective functions and the parameter values, used to compute the Pareto fronts a priori, are needed, as well as the previously calculated Pareto fronts, in order to compute an initial plan. Computing Pareto fronts is performed by the program GAIO [34, 135]. It is also necessary to obtain measurement data which can be used to determine the current working point, in order to be able to compare the model Pareto point with the working point to find out whether or how far they differ from each other. In order to obtain the necessary data, the partial models *Environment*, *Function*, *Behavior*, *Active Structure* and *System of Objectives* from the domain-spanning Conceptual Design Phase (cf. Sect. 2.1) are needed.

3.2.7.2 Description

As mentioned before, the approach described here approximates Pareto fronts, based on selected Pareto points and moving towards a measured working point. This parameter adaptation of a Pareto front must only be carried out if, e. g., environmental changes influence the self-optimizing system to the point that the initial plan becomes invalid. It is possible that an initially computed plan remains feasible during the entire journey of the mechatronic system to its destination. Adapting the parameters is costly, and therefore only beneficial if the initially computed plan is not feasible any longer. This section describes the approach using simple example objective functions for the computation of Pareto fronts.

In order to keep the description simple, we are assuming two-dimensional Pareto fronts with one dimension describing the energy consumption and the other dimension describing the comfort. We assume that a model Pareto point, selected by the planner, and a working point computed using measured values are given. Such a working point can easily be computed if it is possible to measure the current energy consumption and the current comfort.

If these two points are unequal, the model parameters used to compute the model Pareto point deviate from the actual parameters during operation. This can be for several reasons, e. g., due to strong headwind, movement in a convoy or changes of the route. Therefore, the parameters of the multiobjective functions must be re-determined. We distinguish two different cases that allow us to approximate the new Pareto fronts by parameter-fitting.

It is necessary to know both the current model Pareto point being used by the planner and the measured working point. Based on these two points, a part of the model Pareto front close to the selected model Pareto point is selected and recalculated in such a way that it it is shifted towards the working point.

Two approaches are presented here. The first one directly computes new parameter values leading to a Pareto front close to the measured working point. This approach is efficient, but not universally applicable. The second approach uses Taylor series to approximate the Pareto front stepwise and shift it towards the working point.

Parameter Value Determination

If it is possible to determine the changed parameter values, a direct computation is possible. Assume the example objective functions

$$F_{model} = \begin{pmatrix} f_1(ax^2+bx+1) \\ f_2(cx^2+dx+4) \end{pmatrix} \tag{3.20}$$

were used to calculate model data for a Pareto front. Let a, b, c and d be the parameters of the objective functions. The selected Pareto point F_{model} from the Pareto front, computed using a given model, is selected to be $(4,1)$. We assume now that the working point $F_{measured} = (5,2)$ has been determined using measured values from sensors. The variable x is always set to the constant value x_0 and only the variables a, b, c and d can increase or decrease by the values $\Delta a, \Delta b, \Delta c$ and Δd. For this example, we set $x_0 = 1$. The two functions for the model parameter (p) and the measured parameter $\overline{p} = p + \Delta p$ are shown in Equation 3.21 and Equation 3.22. The points used for the description are example points and can take any possible values.

$$F_{model}(x_0, p) = (f_1(x_0, p_1), f_2(x_0, p_2))^T = (4,1) \tag{3.21}$$

$$F_{measured}(x_0, \overline{p}) = (f_1(x_0, \overline{p}_1), f_2(x_0, \overline{p}_2))^T = (5,2) \tag{3.22}$$

To obtain the difference between the Pareto point selected by the planner and the working point calculated using the measured values, the functions can be subtracted from each other. This subtraction leads to the following equations:

$$f_1(x_0, \overline{p}_1) - f_1(x_0, p_1) = 5 - 4$$

$$f_2(x_0, \overline{p}_2) - f_2(x_0, p_2) = 2 - 1$$

Inserting the values from the model and the measured values leads to

$$x_0^2(1+\Delta a) + x_0(2+\Delta b) + 1 - (x_0^2 + 2x_0 + 1) = 1$$

$$x_0^2(1+\Delta c) + x_0(-2+\Delta d) + 4 - (x_0^2 - 2x_0 + 4) = 1.$$

This result shows the difference between the two points in the given dimensions. Simplifying the equations leads to the following notation with the distance between the two points being (1,1).

$$x_0^2 \Delta a + x_0 \Delta b = 1$$

$$x_0^2 \Delta c + x_0 \Delta d = 1.$$

Solving the equations for Δa and Δc gives the notation shown below.

$$\Delta a = \frac{1 - x_0 \cdot \Delta b}{x_0^2}$$

$$\Delta c = \frac{1 - x_0 \cdot \Delta d}{x_0^2}$$

These equations are independent, making it impossible to solve them using Gaussian elimination in order to compute the value changes Δa and Δc. Thus, it is necessary to set Δb and Δd to fixed values in order to be able to solve the equations for Δa and Δc.

As mentioned before, this example describes a case for which it is possible to determine the parameters that have changed. It is assumed that the values a and c have changed. Additionally, it is assumed that $\Delta b = \Delta d = 0$ for the computation of the model Pareto front. This leads to:

$$\Delta a = \frac{1}{x_0^2}$$

$$\Delta c = \frac{1}{x_0^2}.$$

As shown before, x_0 is 1, which leads to $\Delta a = 1$ and $\Delta c = 1$.

It is now possible to compute a new Pareto front using the calculated values for a and c. This Pareto front includes the computed working point, based on the measured values. Based on this new parameter, the Pareto fronts for the upcoming track sections can be calculated and, based on these Pareto fronts, a new plan can be computed.

One disadvantage of this approach is that it must be possible to determine which parameter has changed in the objective functions. If this is impossible, multiple solutions exist. Depending on the assumed values for Δb and Δd, the resulting values for Δa and Δc can be computed. If the assumption is wrong or if the measured working point is far from optimal, the newly computed Pareto front will not be close enough to the measured point to achieve dependability. If the objective functions are infinitely differentiable, another approach can be used to approximate the Pareto front, computed using preliminary parameters close to the measured working point.

Taylor Series Approximation

In a case where it is impossible to determine the changed parameters of the objective functions, an approximation using a Taylor Series expansion can be performed to successively approximate a Pareto front given by model parameters close to a Pareto front obtained using measured values [95].

Assume the differentiable functions, shown in Equation 3.23, which were used to compute a model Pareto front:

$$F_{model} = \begin{pmatrix} f_1(cos(a \cdot x) + b) \\ f_2(sin(c \cdot x) + d) \end{pmatrix} \tag{3.23}$$

Furthermore, assume that the planner has selected a Pareto point at position (2,3) and the measured data revealed a working point at position (6,7). The value of x is always constant. In this example, $x_0 = 1$ is set to a constant value and only the values of the parameters a, b, c, and d are variable. The equations can now be written as follows:

$$F_{model}(x_0, p) = (f_1(x_0, p_1), f_2(x_0, p_2))^T = (2,3)$$

$$F_{measured}(x_0, p + \Delta p) = (f_1(x_0, p_1 + \Delta p_1), f_2(x_0, p_2 + \Delta p_2))^T = (6,7).$$

First, the difference between the two points has to be calculated.

$$f_1(x_0, \overline{p}_1) - f_1(x_0, p_1) = 6 - 2$$

$$f_2(x_0, \overline{p}_2) - f_2(x_0, p_2) = 7 - 3$$

Let x_0 be the Pareto point and let x_m be the working point. To move a Pareto point closer to a working point, each dimension is considered individually. First, a Taylor series for each parameter (p_1, p_2) of the multiobjective function is computed, using the partial derivatives as follows:

$$\begin{aligned} T_i(x) = f(x_0) &+ \frac{\frac{\partial f(x_0)}{\partial x_i}}{1!}(x - x_0) + \frac{\frac{\partial^2 f(x_0)}{\partial x_i}}{2!}(x - x_0)^2 \\ &+ \frac{\frac{\partial^3 f(x_0)}{\partial x_i}}{3!}(x - x_0)^3 + \cdots \end{aligned}$$

Additionally, the differences between the Pareto point and the working point for each parameter $\Delta p_i = \|x_0 - x_m\|_i$ are computed. These differences are used to compute new parameter values p_{i_n}, as shown in the following equation:

$$p_{i_n} = p_i + \frac{\Delta p_i}{T_i(x_m)} : \Delta p_i = \{\Delta p1, \Delta p2\}$$

The Pareto front is then newly computed, based on this new parameters and the procedure is started over again until no further reduction of the differences is achieved. The resulting Pareto front is close to the working point and based on the new environmental parameters. The planner then uses Pareto fronts, based on the newly determined parameters to conduct a replanning, leading to a new and feasible path.

3.2.7.3 Results

As result, a new Pareto front can be computed during runtime, using adapted model parameters. This Pareto front is based on the model parameters which were successively adapted to approximate the model Pareto point towards a measured Pareto point by using the objective functions. Based on the new values obtained for the objective functions, Pareto points for the upcoming track sections can be computed.

These Pareto points can be used to create a new plan using the planner; this new plan will then include current real-world measurements, which leads to an optimized plan for the upcoming track sections and avoids, e. g., an energy consumption that is continuously too high for reaching the goal position. The approach is costly and should therefore only be employed if the existing plan becomes invalid.

3.2.7.4 Application Example

The approach has been implemented and several tests have been performed. A screenshot of the implemented program is shown in Fig. 3.31. It allows the behavior-based adaptation of differing model parameters for various multiobjective functions. The measured working points as well as the model Pareto fronts can be selected as required. The program has also been used to test the combination of the *Behavior-Based Adaptation of Differing Model Parameters* and a planning approach, which is not considererd in this section.

The approach can be applied to single systems (cf. Sect. 1.3.1), as well as to combinations of multiple mechatronic systems for which the resulting settings are based on Pareto points. In general, the approach leads to benefits in all self-optimizing systems that are based on Pareto fronts, if an adaptation of these fronts is necessary, e. g., if the system can be influenced by unpredictable events.

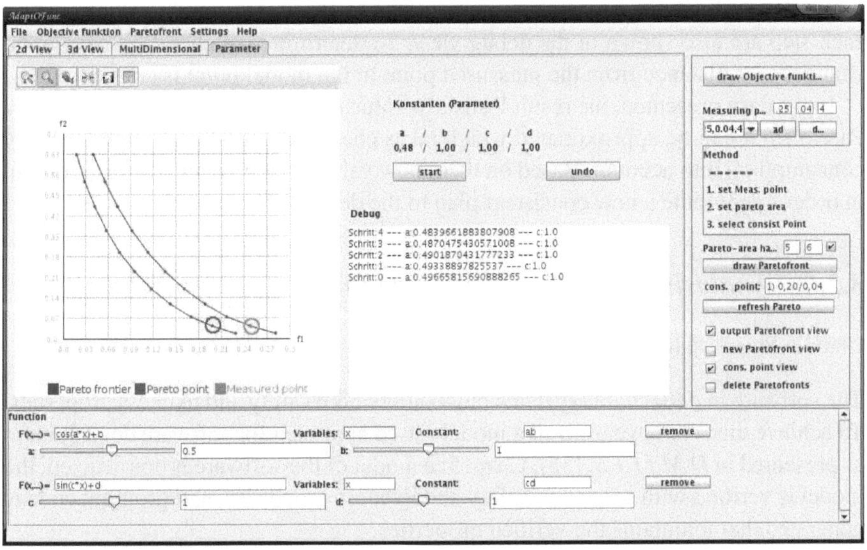

Fig. 3.31 Recalculated Pareto front based on measurement point: the left Pareto front was given by model parameters and the right Pareto front was approximated using the change in the corresponding parameter and the Taylor series expansion. The debug view shows the single parameter settings for the value a to determine the new Pareto front. A measured value as well as the Pareto point selected by a planner can be entered on the right side of the program, for which the Pareto front is then approximated.

Fig. 3.31 shows an example of a recalculated Pareto front based on the Taylor series approach introduced above. The function $f_1 = \cos(a \cdot x) + b$ represents the energy consumption, while the function $f_2 = \sin(c \cdot x) + d$ represents passenger comfort. On the left side of Fig. 3.31, the axis f_1 depicts the current energy consumption and the axis f_2 the current passenger comfort. The planner always receives several Pareto points – in the presented example 11 different Pareto points – from which it has to select one for each track segment. In the depicted example scenario, the planner has selected the Pareto point at position (0.2, 0.04) from the model Pareto front for the current track section (circled point on the left) in order to achieve the required comfort without consuming too much energy. The measured data reveals a current working point at position (0.25, 0.04), visualized by the circled point on the right.

Based on the model Pareto front, a Taylor series expansion is used for stepwise approximation of the given Pareto front close to the working point, obtained from measured values. An approximation is performed in order to be able to compute new Pareto sets online, as the computation of a Pareto front is of high computational complexity. The passenger comfort is still close to the model Pareto point, which means that there must be some model deviation, leading to a higher energy consumption without, however, influencing passenger comfort. To compute the model Pareto front, the values $a = 0.5$ and $b = 1$ were used to calculate the power consumption. One of these variables must differ from the model values, for example because of headwind. As can be see in the debug part of Fig. 3.31, the value a has changed, which moves the complete Pareto front. The single values used in each step are also shown in the debug view. To determine which values have to be changed, the distance from the measured point to the model point is considered.

In the case presented, the result leads to a value $a \approx 0.48$. Using this value, a new Pareto front can be approximated which takes changed values, e. g., higher energy consumption, into account. Based on these new values, the planner is able to re-plan in order to compute a new consistent plan to the destination.

3.2.8 Analysis of Self-healing Operations

Claudia Priesterjahn

The software in dependable systems must satisfy both safety and liveness properties. To achieve this, we have utilized a model-based approach for software development as presented in *D.M.f.I.T.S*, [55], Chap. 5: a model of the software is constructed, the model is verified with respect to safety and liveness properties, and program code is generated that maintains the verified properties.

However, random errors may still occur due to the wear of hardware components, such as sensors, which may affect the software and lead to hazards. *Hazards* are considered situations that "together with other conditions [...] will lead inevitably to an accident" [92]. For example, a speed sensor of the RailCab may fail and lead to the hazard *wrong distance*, which may result in a collision.

Hazards cannot be avoided completely. However, for a dependable system, the developer must guarantee that hazards will only occur with a certain probability. If the occurrence probability of a hazard is too high, the system must be redesigned so that the hazard occurrence probability becomes acceptable [92].

Self-healing (see Sect. 1.1) as a special case of self-optimization may be used to reduce the occurrence probabilities of hazards in mechatronic systems. We propose to use the reconfiguration of the system architecture as a way to stop the propagation of detected failures before a hazard occurs. This is, for example, achieved by disconnecting failed system components and shifting to intact components.

For the specification of failure propagation, we follow the terminology of Laprie et al. [11]. *Failures* are externally visible deviations from the component's desired behavior. They are associated with ports where the component instances interact with their environment. *Errors* are the manifestation of a *fault* in the state of a component, whereas a fault is the cause of an error. Errors are restricted to the internals of hardware nodes.

The errors in hardware components may be detected at runtime using, for example, model-based fault diagnosis [137]. But the error is not observed directly. Rather, the detection will observe a failure at the port of the hardware component or at the port of another component. The errors causing the failure and an appropriate self-healing operation are stored in a fault dictionary [113]. This fault dictionary associates failures in the system with sets of errors causing these failures. The fault dictionary is expanded to include self-healing operations, such that the fault detection not only identifies the causes of a failure, but also triggers a reaction in form of a self-healing operation.

Figure 3.32 shows the architecture of a simplified subsystem of a RailCab that controls the speed of a RailCab in a convoy (see Sect. 1.3.1). The component *sc:SpeedCtrl* represents the speed controller which is responsible for setting the electric current belonging to the linear drive of the rear RailCab in order to have the vehicle drive at a specific speed. The speed controller computes the electric current using the distance between the two RailCabs. This distance is provided by the component *ds:DSelect* which selects the measured distance from two types of sensors, depending on the quality of their data: a distance sensor *dr:DSensor* and a distance computed by *dg:DistGPS*. *dg:DistGPS* computes the distance between the two RailCabs from the position data provided by its gps sensor *gps:GPS* regarding the rear RailCab and the front RailCab. The position data of the adjacent RailCab is provided via wireless network by the components *wlan:WLAN* and *ref:RefData*.

An error in the distance sensor *dr:DSensor* will eventually lead to a failure in the speed controller *sc:SpeedCtrl*. In Fig. 3.32, this error is illustrated by a black circle.

Fig. 3.32 Architecture of the speed control subsystem

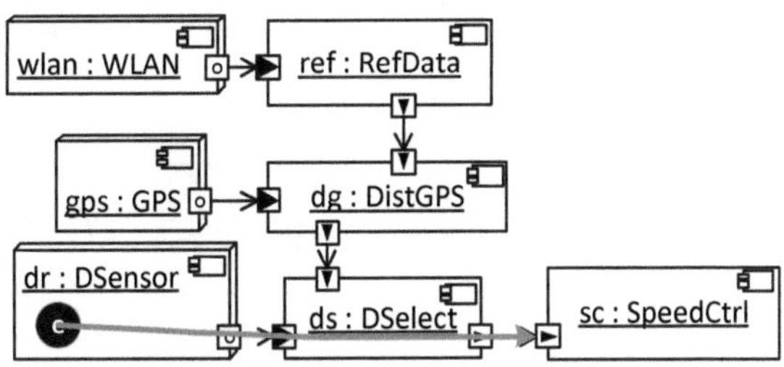

The propagation of the error and the resulting failures is depicted by a gray arrow. The failure of the speed controller causes the hazard *wrong speed*. This means the RailCab will drive at wrong speed, which may result in a collision with another RailCab in the convoy.

Based on the computed hazard occurrence probabilities and the system architecture, the developer constructs a self-healing operation for the hazard *wrong speed*. The self-healing operation is shown in Fig. 3.33; it is triggered when the error $\{e_{dr}\}$ in the distance sensor is detected. The connectors between *ds:DSelect* and *dg:DistGPS*, and *ds:DSelect* and *sc:SpeedCtrl* are removed and, instead, a connector between *dg:DistGPS* and *sc:SpeedCtrl* is created, thereby disconnecting the distance sensor from the subsystem. Distances are then only measured by GPS-based position data.

In order to judge whether such a self-healing operation successfully reduces the occurrence probability of the hazard, we must take into account that the execution of operations as well as the propagation of failures in a real system take a certain amount of time. As a consequence, an analysis of self-healing operations must take the propagation times of failures, the duration of the self-healing operation, and the change that results from the self-healing operation into account. However, current approaches that analyze hazard occurrence probabilities in reconfigurable systems do not consider these properties [59, 64].

Our solution is an analysis of self-healing operations [122, 123] that considers in particular the timing characteristics of failure propagation and the effect of a self-healing operation on the propagation of failures.

The self-healing operations are analyzed when the discrete behavior and the reconfigurations have been specified and verified. The models Environment and Application Scenarios of the Principle Solution (cf. *D.M.f.I.T.S*, [55], Chap. 3) are used to identify hazards. The component structure and behavior models of the MECHATRONICUML (see *D.M.f.I.T.S*, [55], Chap. 5) are used to derive failure propagation models. These models are used to compute hazard occurrence probabilities and to decide which hazard occurrence probabilities need to be reduced via self-healing operations. Then, the self-healing operations are specified and analyzed as described in this section.

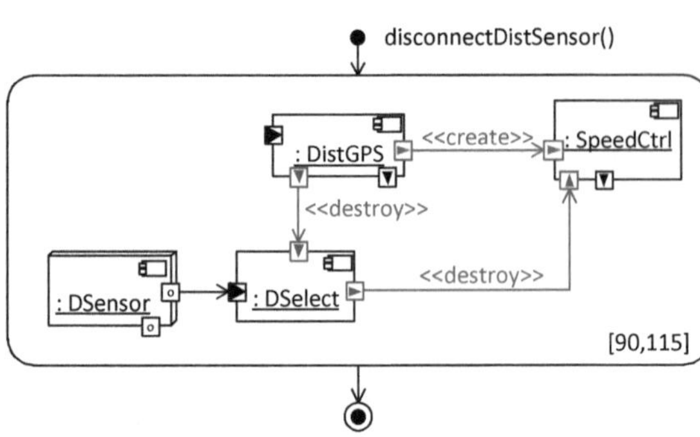

Fig. 3.33 Timed component story diagram for healing the system of a faulty distance sensor

3.2.8.1 Prerequisites and Input

Our analysis of self-healing operations requires information about architecture and behavior of the system, as well as information about the hazards which may occur in the said system.

The architecture and behavior are specified by MECHATRONICUML models (see *D.M.f.I.T.S*, [55], Chap. 5), which have to be constructed manually by the developer. This set of MECHATRONICUML models consists of a deployment diagram specifying the system architecture, the real-time statecharts from the component instances of the deployment diagram specifying the system behavior, and timed component story diagrams (TCSD) specifying the self-healing operations. Additionally, real-time statecharts must be defined for the hardware nodes in the deployment diagram.

The information about the hazard includes a threshold of a hazard occurrence probability for each hazard that may occur. These thresholds are needed to judge whether the self-healing operations reduce hazard occurrence probabilities such that they become acceptable. Further, our analysis requires a set of minimal cut sets (MCS) [92]; these MCSs specify all combinations of errors that could cause the hazard. In order to analyze the reduction of the occurrence probability of a hazard, we need to analyze the self-healing operation for each MCS.

3.2.8.2 Description

Figure 3.34 shows an overview of our analysis of self-healing operations. The required input models were mentioned in Sect 3.2.8.1.

Before our analysis can be applied, the input models need to be created. Our analysis uses timed failure propagation graphs (TFPG); which are failure propagation models with timing annotations. They are generated from the real-time statecharts of the system components.

After creation of the models, a hazard analysis is carried out to compute the occurrence probabilities and MCSs of the system's hazards. Based on the occurrence

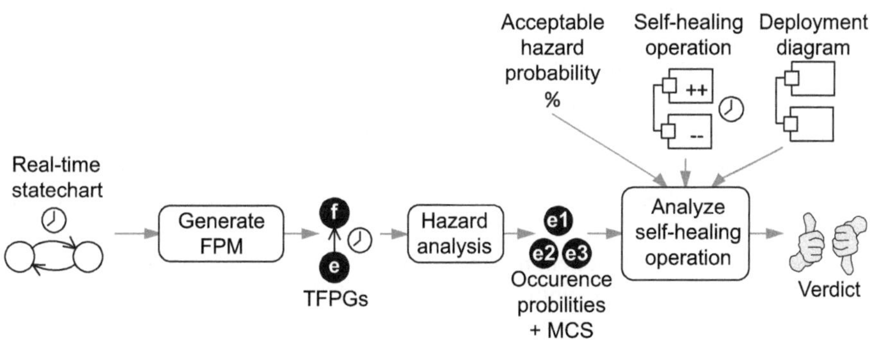

Fig. 3.34 Overview of the analysis of self-healing operations

probabilities, the developer decides which hazards need to be countered by self-healing. For each of these hazards, the developer constructs self-healing operations. This is the only manual step.

Then, each self-healing operation is analyzed. The result is a verdict about the success of the analyzed self-healing operation; it is considered successful if the occurrence probability of the hazard is below the threshold defined in the safety requirements.

Timed Failure Propagation Graphs

Tracing a failure in a system is based on failure propagation models. For analyzing failure propagation times, failure propagation models must include the additional notion of time. Only then it is possible to check whether a self-healing operation has been executed quickly enough. Consequently, our analysis of self-healing operations uses timed failure propagation models called *Timed Failure Propagation Graphs* (TFPG). TFPGs, like common failure propagation models [149], define a cause-and-effect relation between failures. In particular, TFPGs include propagation time intervals that specify minimum and maximum propagation times between failures.

The benefit of TFPGs is their minimal level of information needed for the analysis of failure propagation times. This analysis actually requires taking the reachable behavior of the whole system into account, which comprises the complete data and control flow of the system. However, for analyzing the propagation times of failures, we only need to take the relations between failures at the ports of components into account. We therefore analyze the reachable behavior of each component type (see *D.M.f.I.T.S*, [55], Chap. 5) only once to identify these relations and store them in a TFPG. During our analysis of self-healing operations, we abstract from the system behavior using TFPGs.

Failures are classified using a failure classification like the one by Fenelon et al. [47]. We distinguish the failure classes *value*, *service*, *early timing*, and *late timing*. A value failure specifies a deviation from a correct value, e.g., an erroneous parameter of a message. A service failure specifies that no value is present at all, e.g., a component crashed and is not providing any output values. A timing failure specifies that a message has been delivered outside a defined time interval, i.e., too early or too late.

Figure 3.35 shows the TFPG of the deployment diagram from Fig. 3.32. In TFPGs, failures are represented by rectangles labeled with the according failure. Operators are represented by circles labeled with the according logical operator. Edges are labeled with propagation time intervals that specify the minimum and maximum propagation time that a failure needs to propagate from the edge's source to the edge's target.

The TFPG of the component instance *ds:DSelect* from Fig. 3.35 relates the outgoing failure f^o_{ds3} to the incoming failures f^i_{ds1} and f^i_{ds2}. The operator *OR* specifies that f^o_{ds3} occurs if either of the incoming failures occurs. The failures of the TFPG of *ds:DSelect* are connected to failures of connected component instances according to the connectors of the deployment. For example, the outgoing value failure f^o_{ds3} of *ds:DSelect* is connected to the incoming failure f^i_{sc1} of the speed controller, because *ds:DSelect* and *sc:SpeedCtrl* are connected. The edge is labeled with the propagation time interval $[5,6]$ of the connector. This edge specifies that the propagation of a failure from *ds:DSelect* to the distance controller takes between 5 and 6 time units.

The propagation time interval at the edge from f^i_{ds2} to the OR-node (see Fig. 3.35) specifies that a failure needs at minimum 24 and at maximum 28 time units to propagate from the port of the component instance *ds:DSelect* to the OR-node. The edge originating from the OR-node has a propagation time interval of $[0,0]$. This means that between the OR-node and the outgoing port, failures propagate in zero time. These time intervals are introduced by automatic generation. Thus, a failure needs between 24 and 28 time units to propagate from the incoming ports of *ds:DSelect* to the outgoing port.

The edges connecting the errors in the hardware components and their outgoing failures have relatively high propagation times compared to real-life sensors. This is because we have adjusted these times to reflect the fact that sensor failures may not have an immediate negative impact on the system. If, for example, a sensor delivers a single peak value, this peak value may be corrected by smoothing the signal using a low-pass filter. It is only when the deviation from the correct signal occurs in a number of subsequent signals that this is interpreted as possibly having an adverse effect on the system. This fact needs to be taken into account during the analysis of the self-healing operations because there is more time for the system to react by self-healing if a controller tolerates a certain amount of deviating values. We call this time *tolerance time*.

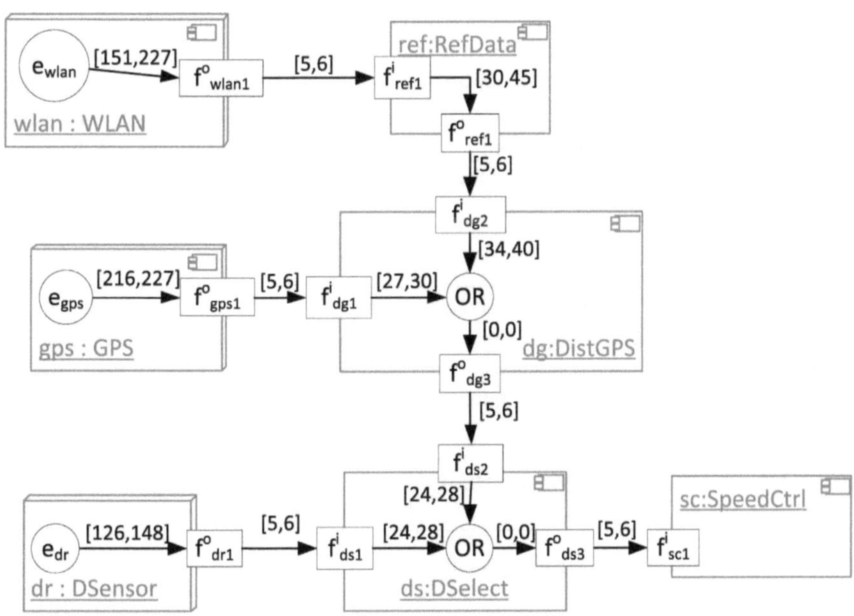

Fig. 3.35 Timed failure propagation graph of the deployment diagram from Fig. 3.32

In order to take this fact into account in our analysis of self-healing operations, we include the tolerance time in the TFPG. The tolerance time is computed from the product of the data rates of sensors and the number of repeated deviations the controller is able to tolerate. The tolerance time is added to the propagation times that errors need to propagate to cause a failure at the ports of the sensors.

In order to analyze how far failures propagate within a given time span, we need to define a semantics for TFPGs. Therefore, we map TFPGs onto *Time Petri Nets* (TPN) [129]. TPNs are marked petri nets [129] with a time extension. Each transition in the TPN has an interval that specifies its earliest and latest firing time. The propagation of failures over time is then analyzed by the reachability analysis of TPNs espoused by Cassez et al. [26].

Generation of Timed Failure Propagation Graphs

The goal of our TFPG generation is to compute relations and propagation times between incoming and outgoing failures of a particular component type. Figure 3.36 shows an overview of our TFPG generation as published in [119]. The input is the real-time statechart that specifies the behavior of the component type.

In order to construct the TFPGs, we first identify which incoming failures cause which outgoing failures for each component type. Each time such a relation is identified, the propagation times between the related failures are computed and the relation is stored in a TFPG. This is repeated until all combinations of incoming failures have been evaluated.

We must distinguish between the identification of relationships between outgoing and incoming timing and service failures on one hand, and outgoing and incoming value failures on the other hand. Service and timing failures change the control flow such that either no message is sent, or a message is sent too early/too late. Causes for these deviations may be that either transitions which should have been fired could not be activated or that other transitions with other time constraints have fired. Relations between incoming and outgoing timing and service failures are therefore identified by deviations in the control flow.

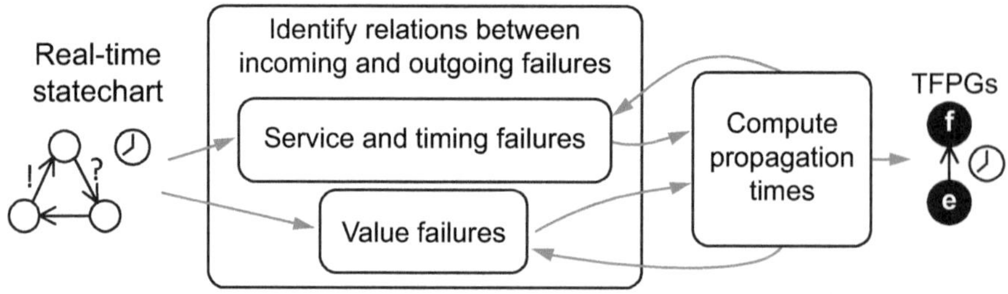

Fig. 3.36 TFPG generation for a component type

To provoke these deviations, the control flow is modified by injecting failures into the real-time statechart. Failures enter a component via faulty messages; thus, to inject a timing failure, a message is sent earlier or later than specified by the real-time statechart. For a service failure, a message expected by the real-time statechart is not sent at all. These modified messages may change the control flow of the real-time statechart, allowing outgoing timing and service failures to be identified by deviations between the original and the modified control flow.

Value failures cannot be detected by deviations in the control flow, because even though the same transition is fired, the values of variables may differ. Consequently, we need to identify relations between incoming and outgoing value failures from the data flow.

Identifying relations between incoming and outgoing value failures is based on the slicing of extended finite state machines shown by Androutsopoulos et al. [8]. It is the only approach for slicing nonterminating automata, which makes it the only approach suitable for embedded real-time systems.

To apply this slicing, we map real-time statecharts onto extended finite state machines. The resulting slice is an extended finite state machine that contains only the parts of the real-time statechart which affect a specific variable. To identify which incoming value failures cause outgoing value failures, we compute the slice of each variable v sent by the real-time statechart as a message parameter. The remaining variable assignments are those which influence v or are influenced by v.

Analysis of Self-healing Operations

Our analysis of self-healing operations as published in [122, 123] checks for each MCS of the hazard, whether the MCSs can still cause the said hazard after the self-healing operation has been completed. MCSs that still cause the hazard after self-healing are called *critical MCSs*. If the number of MCSs of the hazard is reduced, the number of events that cause the hazard will be reduced as well. As a consequence, the occurrence probability of the hazard is decreased.

The TFPGs which provide the input for our analysis of self-healing operations have been generated from the real-time statecharts of system components, as explained above. The TFPGs are used to compute the MCSs and the hazard occurrence probabilities by the component-based hazard analysis of Giese et al. [59]. Therefore during this analysis, the timing annotations of the TFPGs are ignored.

Based on the computed hazard occurrence probabilities and the system architecture, the developer constructs a self-healing operation for the hazard. Based on the real-time statecharts, our analysis computes the *critical time*. The critical time is the maximum amount of time between the detection of the error or failure and the last point in time when the self-healing operation can successfully be executed. This is the time span during which a failure will propagate through the system before the self-healing operation has been completed, and may vary within a certain interval due to system-specific properties. We take the maximum value of this interval in order to analyze the worst case: the failures propagate as far as possible.

Next, our analysis computes how far the errors of the MCS have propagated through the system during the critical time. The result are the errors and failures which are reachable before the execution of the self-healing operation.

Then, the self-healing operation is applied. It changes the structure of the deployment and thereby the structure of the TFPG. This may cut off propagation paths along which failures propagate to the hazard, or may remove errors and failures from the system. The result of this step are the errors and failures that remain in the system after the application of the self-healing operation.

In the next step, the criticality of the MCS is evaluated based on the errors and failures which remain in the system meaning that the analysis checks whether the errors and failures which remain in the system after the application of the self-healing operation still lead to the hazard. If this is the case, the MCS is critical.

After the criticality of all MCSs has been analyzed, the occurrence probability of the hazard is computed based on the critical MCSs. Finally, our analysis checks whether this computed occurrence probability is acceptable. If it is acceptable, the self-healing operation has been successful in reducing the hazard. Otherwise, the self-healing operation has failed and the developer has to either modify the self-healing operation or utilize another technique to reduce the hazard occurrence probability.

3.2.8.3 Results

The analysis of self-healing operations is used to implement a system such that all hazard occurrence probabilities are acceptable. This analysis is used in two ways: It enables the developer to either guarantee that the self-healing operations will reduce the occurrence probabilities to an acceptable level or the analysis shows that the resulting hazard occurrence probabilities are not acceptable. In the latter case, the developer will change the system and analyze it again. This change may affect all preceding steps of the development process which have already been carried out before.

3.2.8.4 Application Example

In our example, the result of the component-based hazard analysis are the minimal cut sets $\{e_{wlan}\}$, $\{e_{gps}\}$, and $\{e_{dr}\}$, because any of the errors in the hardware components of the speed control subsystem may cause the hazard. The occurrence probability of each of the errors is 0.001 leading to the occurrence probability of the hazard of 0.003.

Based on the computed hazard occurrence probabilities and the system architecture, the developer constructs a self-healing operation for the hazard *wrong speed*. The self-healing operation is shown in Fig. 3.33.

Next, our analysis computes how far the errors of the MCS have propagated through the system during the critical time. In our example, we compute a critical time of $[120, 140]$. This means the amount of time between the detection of $\{e_{dr}\}$

and the completion of the self-healing operation lies between 120 and 140 time units.

Then, the self-healing operation is applied. Figure 3.37 shows the TFPG of the speed control subsystem after the application of the self-healing operation. The failures, which are reachable within the critical time, are highlighted in gray.

The self-healing operation of our example in Fig. 3.33 removes the connectors between *dg:DistGPS* and *ds:DSelect* and between *ds:DSelect* and *sc:SpeedCtrl*. It further creates a connector between *dg:DistGPS* and *sc:SpeedCtrl*. By deleting the connectors, the corresponding edges in the TFPG are deleted, as well. Consequently, the edges between f^o_{dg3} and f^i_{ds2} and between f^o_{ds3} and f^i_{sc1} are removed.

In our example, none of the errors and failures which remain in the system after the application of the self-healing operation can propagate to the speed controller. The MCS $\{e_{dr}\}$ is consequently not critical.

The critical MCSs after the self-healing operation are $\{e_{wlan}\}$ and $\{e_{gps}\}$. These errors may still cause the hazard.

The reduced occurrence probability of the hazard is $p(e_{gps}) \vee p(e_{dr}) = 1 - ((1 - p(e_{gps}))(1 - p(e_{dr}))) = 1 - ((1 - 0.0001)(1 - 0.0001)) = 0.0002$. Thus, the occurrence probability of the hazard has been reduced below the maximum acceptable occurrence probability of 0.001.

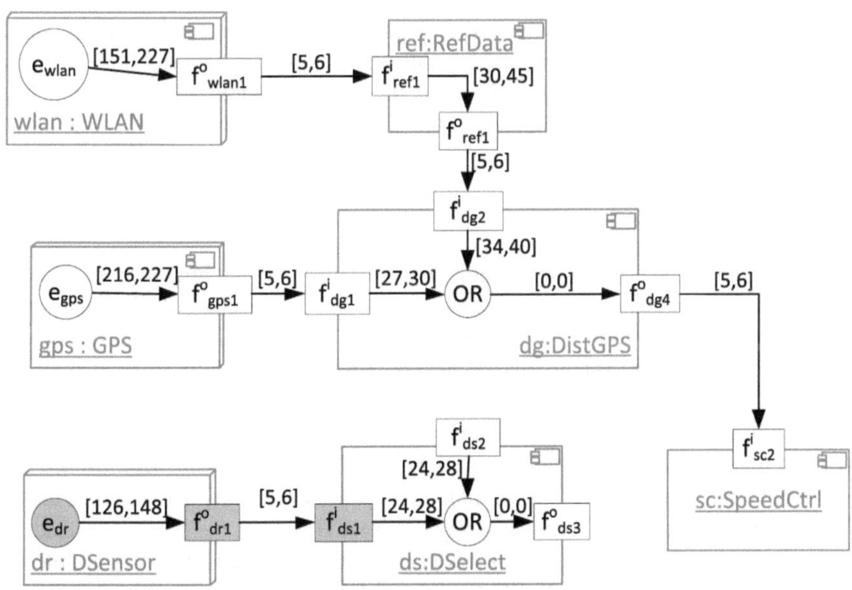

Fig. 3.37 TFPG after the reachability analysis and the application of self-healing operation

3.2.8.5 Further Reading

Consistency

The architecture of the system is provided by the Conceptual Design Phase (see Sect. 3.1) in order to guarantee a consistent transition into Design and Development. For our analysis of self-healing operations, these architecture models are refined and expanded to include real-time statecharts and reconfiguration behavior. The system model is changed if the occurrence probability of the hazard is still too high after self-healing. These changes are fed back into the models of the Conceptual Design Phase. This scenario was carried out in the student project "SafeBots II" [9].

Hazard Analysis for the Entire Mechatronic System

For mechatronic systems, uniting four disciplines in one system requires the development and analysis of the system as a whole. The key difference between this process and pure software architecture is that (hardware) connectors that are connected to hardware components transport information but also physical items, i.e., material and energy. The partial model Active Structure (see *D.M.f.I.T.S*, [55], Chap. 4) specifies the architecture of the entire mechatronic system, including both hardware and software. Hardware connectors are only represented as simple connections, even though they may correspond to additional system components. In [121], we present a component-based hazard analysis that considers the whole mechatronic system, including hardware connectors, and that introduces reusable patterns for the failure behavior of hardware connectors which can be generated automatically. In this way, the component-based hazard analysis of Giese et al. [59] can be applied to the entire mechatronic system.

Runtime Hazard Analysis

In self-optimizing systems, it is possible that certain system architectures may occur only at runtime and cannot be foreseen at design time. Consequently, in the domain of self-optimizing systems, not all system architectures can be analyzed at design time. In order to still be able to guarantee certain hazard probabilities for the system, hazard analysis needs to be performed at runtime.

When, for example, RailCabs become ready for the market, they will be produced by more than one source. In such a case, it is possible that two vehicles must interact which have been produced by different manufacturers. Of course, it would be useful for both vehicles to form a convoy to save energy. However, the developer needs to analyze whether the communication between the two RailCabs does not violate the safety requirements of either RailCab. This analysis may not have been possible at design time, because the developer of one RailCab may not have known the system models of the other RailCab. Such a scenario is one reason to compute the hazard probabilities of a system at runtime, in order to guarantee safety requirements for component architectures unknown at design time, as presented in [120, 124].

3.2.9 Safe Planning

Steffen Ziegert

The goal of this method is to provide the self-optimizing mechatronic system with a means of making decisions autonomously about the application of reconfigurations which affect the system's architecture. This is an important ability of the system in connection with the dependability attribute "availability". By considering safety requirements when making its decisions, this method also meets requirements of the dependability attribute "safety". The method is applied in the software engineering domain and assumes a prior verification of the system's communication behavior and reconfiguration operations with the method presented in *D.M.f.I.T.S*, [55], Sect. 5.2.3.2.

Adapting to a new situation calls for a number of runtime reconfigurations that may include changes to the system's architecture, such as the creation and deletion of component instances or communication links between them. For each configuration of the system, there may be a large set of applicable runtime reconfigurations. Selecting which runtime reconfigurations to apply can be a complex task. Self-optimizing systems often have superordinated goals that are supposed to be reached during operation, such as optimizing the energy consumption or achieving user-specified objectives. These goals have to be taken into account when selecting which runtime reconfigurations to apply. However, selecting runtime reconfigurations that are likely to help in achieving the goal is no trivial task. Since the selection of runtime reconfigurations is intended to be autonomous (human intervention would not meet the response-time requirements of self-optimizing mechatronic systems), it has to be planned by a software system.

To prevent unsafe configurations, e.g. an unsafe distance between two RailCabs, from occurring in a plan, the planning system should further take safety requirements into account. The safety requirements restrict the set of valid configurations, i.e. they specify which configurations are not allowed to occur in a resulting plan. In contrast to the design-time verification, where the absence of unsafe states is categorically guaranteed, this technique allows unsafe states to exist in the reachability graph, but plans the reconfigurations in such a way that no unsafe state is actually reached.

3.2.9.1 Prerequisites and Input

For this method to be applicable, a set of possible runtime reconfigurations has to be specified beforehand. This in turn necessitates both the specification of the system's structure as a MECHATRONICUML component model (cf. *D.M.f.I.T.S*, [55], Sect. 5.2) and the mutual comparison of this component model with the models used by other methods, encompassing the agreement on a common model for the rail system. This specification is based on the partial models "Active Structure" for the system's structure and "Behavior – Activities" for the runtime reconfigurations of the system.

To meet the safety requirements, a set of forbidden patterns, i.e. the specification of subconfigurations that are not allowed to appear in a safe plan, is required. Furthermore, this method requires either a set of initial configurations given as MECHATRONICUML component instance configurations when used for simulation or the capability of perceiving the current configuration of the system when used online.

3.2.9.2 Description

Depending on the application domain, it can be very complicated to guarantee the absense of all forbidden patterns by means of design-time verification (cf. *D.M.f.I. T.S*, [55], Sect. 5.2.3.2). The exclusion of a forbidden pattern via design-time verification imposes restrictions on the state space of the system, as it is not allowed to contain states matching the forbidden pattern. This, in turn, transfers these restrictions to the design of the application domain's reconfigurations. If these restrictions prove too cumbersome, we can instead allow the forbidden patterns to appear in the state space in principle, but plan such that they do not appear on the path to the target configuration. Of course, when applying the method *Safe Planning*, we do not need to take forbidden patterns into account whose absense has already been verified by the design-time verification.

Problem

Our approach uses graph transformation systems as the underlying formalism. A graph transformation system consists of a graph representing the initial configuration of the system, plus a set of graph transformation rules (GT rules). These rules schematically define how the graph can be transformed into new configurations by means of two graphs, called left-hand side (LHS) and right-hand side (RHS), and a morphism between them. The morphism identifies the objects and links that are preserved when the GT rule is applied. Other elements specified in the LHS and RHS are deleted and created, respectively, when the GT rule is applied. Syntactically, such a rule can be represented by a story pattern that integrates the LHS and RHS into one graph by using stereotypes [51].

To apply a GT rule to a host graph, a match from its LHS to the host graph first has to be found. This match defines the subgraph of the host graph being manipulated by the rule application. Note that by finding multiple matches in the host graph, one GT rule can spawn multiple graph transformations. In other terms, the GT rule can be seen as a *parameterized* action and the graph transformation can be seen as a *grounded* action in which the elements of the LHS have been substituted with the elements from the host graph.

The transition system of the graph transformation system can be constructed by successively applying the graph transformations to the initial configuration and its successor configurations. The planning task is to find a path in this transition system ending in a target configuration. A safe planning task is principally the same, but includes the requirement that no potentially unsafe configuration is reached.

We can now give a general definition of the safe planning problem and relate it to our modeling approach:

Definition 3.1. A *safe planning problem* is a tuple $P = (I, A, G, F)$ where

- I is the initial configuration,
- A is a set of (parameterized) actions,
- G is a goal specification that defines whether a configuration is a target configuration, and
- F is a safety specification that defines whether a configuration meets the safety requirements and is allowed in a safe plan.

In our case, the initial configuration corresponds to a UML object diagram that has been derived from a MECHATRONICUML component instance configuration. Each parameterized action is given as a story pattern. For a given configuration, the set of successor configurations can be computed by matching the story patterns' LHS's to the host graph and applying the graph transformations that are defined in doing so.

The goal specification is a function that maps from the set of states to booleans. Within the context of this method, the goal specification is given as a graph pattern, i.e. a story pattern without a RHS. Each host graph that the pattern matches onto is a target configuration.

The safety specification is also a function that maps from the set of states to booleans. It is given as a set of graph patterns, called forbidden patterns. If any one of the forbidden patterns matches the host graph, the host graph is a forbidden configuration that does not meet the safety requirements. In contrast to the goal specification, the safety specification does not map to *true*, but rather to *false* when *any* pattern matches.

Solution

Different algorithms and techniques exist to solve these planning tasks. One of the approaches is to translate the planning problem into an input for available off-the-shelf planning system. These traditional planning systems, however, work on models which are different from graph transformation systems. They are used for models with first-order literals that are usually compiled into a propositional representation by grounding predicates and actions. While a translation is technically possible, there are some restrictions because typical planning languages, such as the Planning Domain Definition Language (PDDL), which is the current *de facto* standard in academia, have a different expressive power than graph transformation systems. Today's proposed translation schemes [100, 144] do not support arbitrary negative application conditions and cannot work with an unlimited number of objects. For instance, a planning model designer would have to specify the maximum number of objects beforehand. Howver, this number is not necessarily finite. By planning directly within the transition system defined by the graph transformation system, we avoid these problems.

For planning with graph transformations, we have developed two approaches that complement each other. The first approach diverts a model checker from its intended use of searching for a counterexample of a given property [132]. It plans by reformulating the planning problem into a model-checking problem and then asking a model checker to verify the property that *no* plan exists. If the property is false, i.e. a plan exists, the model checker delivers a plan as counter example of the property. While this approach is very generic and fully automatic, it is not competitive in terms of speed and quality compared to other planning techniques because the state space search of a model checker is generally not optimized for planning, i.e. for finding a plan quickly or for finding a short plan.

Our second approach is based on heuristic search. We use search algorithms like A* or Best First (cf. [134]) with a domain-specific heuristic to search through the state space. The advantage of this approach is at the same time its disadvantage: on the plus side, this planning technique uses guided search, which is in general faster than the model-checking-based approach; however, the disadvantage is that a heuristic suitable for the given application domain first has to be developed. A solution to this problem is to learn heuristic functions automatically [44]. A learning algorithm derives a *regression function* that predicts the costs of solving the problem from a given state. To derive the regression function, the learning algorithm needs a pre-defined declaration of state features and a training set with problem instances. While this solution is an improvement over developing heuristic functions manually, it still requires the developer to declare a set of state features that is suitable for the given application domain. To overcome this issue, we are currently investigating the use of domain-independent heuristics for planning with graph transformations.

Both approaches support forbidden patterns and thus are capable of solving the safe planning problem. In the model-checking-based approach, the problem is solved by including the safety requirements into the property to be verified. Now the property states that no *safe* plan exists, i.e. there is no path to a goal state free of intermediate (on-the-way) states containing forbidden patterns. Thus, the model checker must also consider whether any state on the path to the goal state contains a forbidden pattern. In the heuristic search-based approach, checking for forbidden patterns is simply integrated into the search algorithm.

Translations into dedicated planning languages do not support forbidden patterns at the moment. Although planning languages which support constraints exist, translation schemes do not yet support the translation of forbidden patterns into such constraints. Proposed translation schemes [100, 144] only support the translation of GT rules into the action representations known from PDDL. As remarked above, their capability to support negative application conditions is also limited.

Which one of the two approaches to planning with graph transformations is preferable, depends on the application domain. If the application domain allows for a straightforward design of a suitable heuristic using human intuition, or provides a meaningful set of state features to derive a heuristic function using machine learning techniques, then the heuristic search is to be preferred. If, however, the domain does not provide a meaningful set of state features, meaning that heuristic knowledge is not easy to obtain, then the model-checking-based planner might be the better choice.

3.2.9.3 Results

The result of the integration of this technique into the Operator Controller Module (cf. Sect. 1.1.1) is its ability to autonomously plan which runtime reconfigurations to execute for a given state of the system. The planning process itself is performed by the Cognitive Operator of the OCM. The outcome of this process is a plan, i.e. a sequence of graph transformations, to reach a target configuration. These resulting runtime reconfigurations are executed by the Reflective Operator of the OCM.

Since the planning process considers safety requirements in the form of forbidden patterns, the Reflective Operator is guaranteed not to execute any runtime reconfigurations leading to a dangerous situation.

3.2.9.4 Application Example

A *real-time coordination pattern* represents a communication protocol between subsystems that is specified by real-time statecharts (cf. *D.M.f.I.T.S*, [55], Sect. 5.2.1). We use the activation and deactivation of real-time coordination patterns in the RailCab system (cf. Sect. 1.3.1) as an Application Scenario. In this scenario, the railway network consists of a set of tracks that are connected via successor links. Each track is further monitored by one or more base stations. A RailCab has to register itself at such a base station in order to continuously provide information about its exact position and enable itself to request information about the properties of the track segment (cf. *D.M.f.I.T.S*, [55], Sect. 2.1.6).

An initial configuration is given in Fig. 3.38. It shows two RailCabs on a track that belongs to the station *Paderborn* and a railway network that connects this station to the stations *Berlin* and *Leipzig*.

The story patterns specify the runtime reconfigurations that allow the transformation of this configuration into the target configuration. One of these story patterns, the story pattern *cPublication*, is shown in Fig. 3.39.

It specifies the creation of a real-time coordination pattern between a RailCab and a base station under the condition that the RailCab is occupying a track segment monitored by the base station. For unregistering a RailCab there is a similar story pattern called *dPublication*. In addition to the story patterns for registering and unregistering a RailCab at a base station, there are story patterns for moving a RailCab, initiating the formation of dissolving of a convoy of RailCabs, joining and leaving a convoy, and moving a convoy.

One of the safety requirements for this application scenario states that a RailCab may not operate in a convoy if it holds dangerous cargo. This requirement is formalized by the forbidden pattern *dangerInConvoy* which is shown in Fig. 3.40. Other requirements state that there may be no collision or a distance small enough to be unsafe between two RailCabs, and that a RailCab may not be registered at an

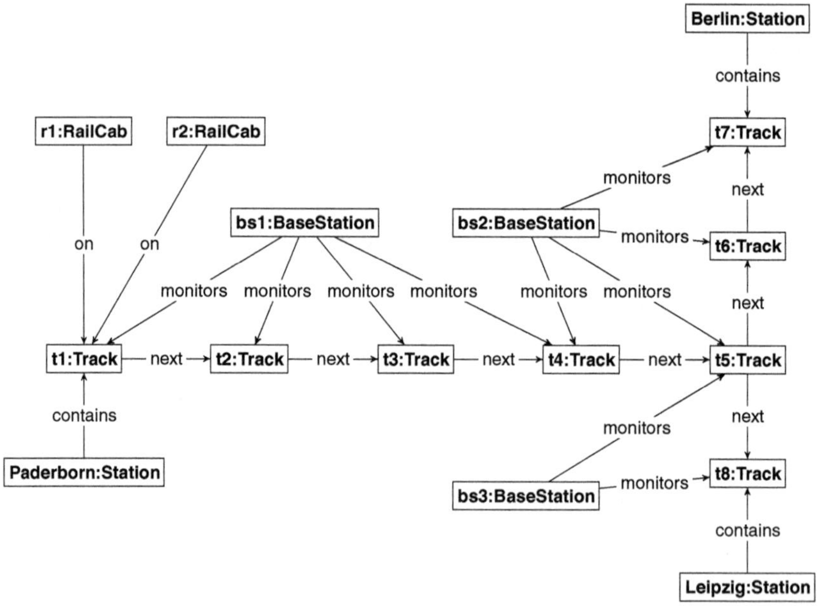

Fig. 3.38 An initial configuration given as a UML object diagram derived from a MECHATRONICUML component instance configuration

incorrect base station, i.e. a base station that does not monitor the track segment the RailCab is occupying.

Given a goal specification, e.g. a graph pattern which states that the RailCab *r1* is occupying a track segment connected to *Berlin* and *r2* is occupying a track segment connected to *Leipzig*, this model can be fed into one of the planning systems introduced above. The planning system then directly plans in the transition system corresponding to the given model. Therefore, no translation to a dedicated planning language and thus no restriction on the expressive power of graph transformation systems is necessary. Unsafe configurations are recognized by the planning system and not allowed in a valid plan. The resulting plan specifies a sequence of runtime reconfigurations that safely turn the system from its initial configuration into a target configuration.

Fig. 3.39 A story pattern specifying the instantiation of a real-time coordination pattern for the communication between a RailCab and a base station

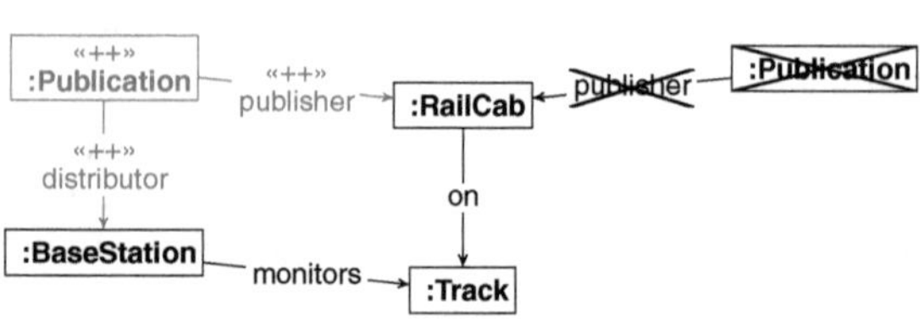

Fig. 3.40 A forbidden pattern that prohibits a RailCab's inclusion in a convoy if it holds dangerous cargo

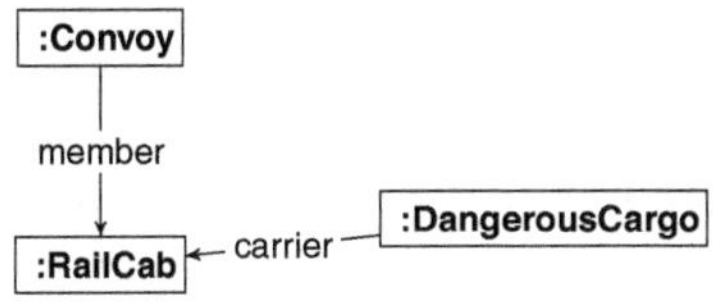

3.2.10 Verification for Interacting Mechatronic Systems with Motion Profiles

Kathrin Flaßkamp, Christian Heinzemann, Martin Krüger, Sina Ober-Blöbaum, Wilhelm Schäfer, Dominik Steenken, Ansgar Trächtler, and Heike Wehrheim

The method is intended to verify the correctness of the behavior of interacting self-optimizing mechatronic systems in specific Application Scenarios. Verification methods play a crucial role in ensuring dependability, most importantly the attribute "safety", in a technical system. In this section, we present an offline verification method which is applied during the development phase of the system. It requires input from several partial models of the Principle Solution (cf. Sect. 2.1), as the dynamical (time-discrete and time-continuous) behavior can be derived from the partial models "Behavior" and "Active Structure", in some cases in combination with the "Environment" model. Certain details of our verification method have to be adapted to the specific scenario chosen from the list of relevant scenarios stored in the partial model "Application Scenarios"; for example, we will use a specific convoy braking scenario in the following. Motion profiles which abstract the continuous time behavior can be computed via multiobjective optimization to take into account objectives from the partial model "System of Objectives".

A central aspect of mechatronic systems is their hybrid nature, i.e. they have continuous (formalized, e.g., by differential equations), as well as discrete (formalized, e.g., by timed automata) behavior. Most state-of-the-art modeling approaches which deal with such hybrid systems use formalisms such as hybrid automata [69]. In doing so, they move into areas of complexity that make verification impossible or at least unfeasible, except for tiny "toy" examples [5]. In contrast, our approach relies on so-called *motion profiles* that fulfill a given set of properties. A motion profile is essentially a curve or set of curves that shows the development over time of physical parameters describing the motion of the vehicle. These motion profiles can be generated with simulation and optimization methods which use models of the system's dynamics. The distribution of these motion profiles can be modeled with (timed) discrete formalisms that can be verified with much less effort. Correctness is implied if the distribution satisfies certain constraints, based on the properties of the profiles.

The verification of a behavior specification based on motion profiles seems especially suitable for interacting mechatronic systems. Classically, the dynamics of each individual system is controlled by a feedback / feedforward control strategy based on a model of the time-continuous dynamics. In addition to this, the interaction of the systems is specified by a communication protocol in terms of discrete automata that have to be checked for correctness using formal verification techniques. For illustration, we will continuously refer to the example of a convoy of RailCabs (cf. Sect. 1.3.1) throughout this section.

The method described in this section provides the user with tools and processes that allow the correct implementation of motion profile creation, distribution and usage. There are two substantially different ways to approach profile creation and distribution: online generation and offline generation. While we will mainly focus on offline generation in this text, we will also give a short overview of online generation.

In offline generation, each RailCab is equipped at the time of design with a set of motion profiles for every maneuver it will need to perform in the future. Based on analytical functions comparing two motion profiles, a communication protocol distributes the available motion profiles and determines which of these motion profiles is active for any given maneuver at runtime. Both the analytical functions and the communication protocol can be verified at design time. The following subsections will go into some detail of offline profile generation, the analytical functions, the communication protocol and its verification.

In contrast, for online generation, the RailCab is only equipped with some basic physical information about itself at design time. At runtime, motion profiles are generated on the fly when the situation is about to change, for instance when the RailCab is about to enter a convoy. From environmental data and the physical data about the RailCab, new instances of prepared constraints are derived for motion profiles for each maneuver that will be affected, and profiles are constructed that fit these constraints. Once all profiles are complete, the RailCab can proceed; it will either enter the convoy, or will be forced to abort if one of the computations fails to come up with a valid profile in time.

3.2.10.1 Application Example

The example we will use to illustrate this method is convoy maintenance; the particular situation under consideration is that a RailCab wants to join an existing convoy (see Fig. 3.41). In such a situation, multiple motion profiles have to be chosen and exchanged; for the sake of simplicity, however, we will focus on only one of them: controlled braking. This motion profile is responsible for the controlled stop at a train station or switch. In convoy mode, it is of great importance that the braking profiles present in the convoy are compatible with each other; if one of the RailCabs were to decelerate faster than the RailCab following it, they would collide.

Thus, for a candidate position in the convoy, the joining RailCab must choose a profile that is compatible (as described in *Profile Compatibility*, Sect. 3.2.10.3) with the profile of the RailCab preceding it and the profile of the RailCab following it. From our point of view, this means that the RailCab must choose a profile for which the analytical function mentioned above indicates that it is compatible with the two motion profiles concerned. If there is no such profile for any position in the convoy, then the RailCab cannot safely enter the convoy and must abort the process.

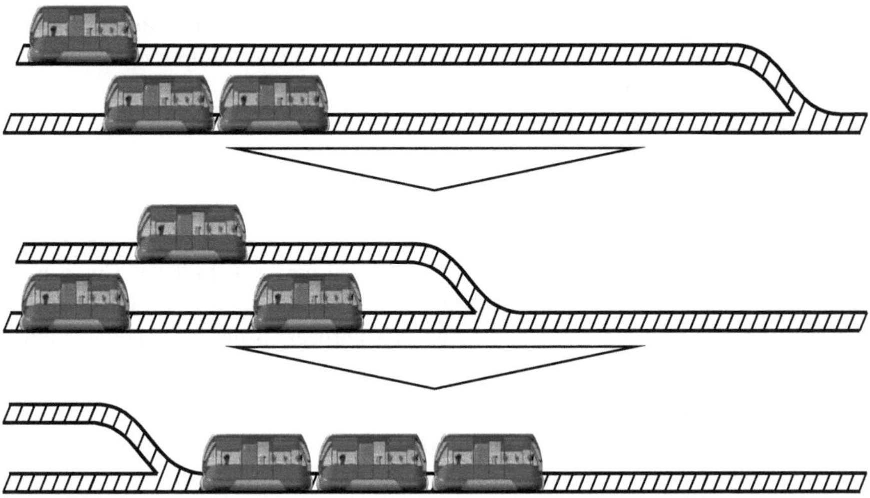

Fig. 3.41 The convoy merging scenario

In all, three components influence the correctness of the process that assigns motion profiles: the selection of profiles with which a RailCab should be equipped, the analytical function used for comparing profiles and the communication protocol used for distributing profiles in a convoy.

The crucial steps of the method involving offline profile generation are the following:

1. generation of profiles,
2. design of the communication protocol for distributing the profiles,
3. verification of the communication protocol,
4. modeling of the compatibility function,
5. checking the profile compatibility, and
6. distribution of profiles.

Steps 1-4 are performed at design time, while Steps 5 and 6 are performed at runtime. We will touch on each of these subjects in turn in the following sections.

3.2.10.2 Prerequisites and Input

For offline profile generation, the following pieces of information are required at design time:

- The full range of the physical parameters of the RailCab relevant to its dynamics, such as mass, external dimensions, motor strength, etc., together with a model that describes its continuous time dynamics
- A description of every behavior (maneuver) the RailCab may engage in during runtime, such as entering convoys, leaving convoys, allowing others to enter, stopping at train stations, etc.

- A mathematical formulation of the set of objectives related to these different behaviors, such as, for example, minimizing energy consumption, minimizing travel time or maximizing passenger comfort

3.2.10.3 Description

Profile Generation

To begin with, a set of profiles has to be generated for each RailCab. This can be done using model-based optimization techniques, such as optimal control, for the computation of optimal braking maneuvers. This process is dependent on a number of different parameters that can be categorized as

constant parameters:
: These parameters do not vary during operation of the RailCab. An example of such a parameter is the physical shape of the RailCab.

discrete parameters:
: These parameters span a finite set of values – each combination of such parameters necessitates at least one motion profile per maneuver. One such instance of discrete partameters could be a predefined (small) number of different cruising speeds of convoys.

continuous physical parameters:
: These parameters change from one operation to the next and span a subset of the real numbers, an example of this type being the RailCab's mass. They have to be discretized into regions.

objectives' priority parameters:
: These continuous parameters weight the objectives and may vary during operation of a self-optimizing system. For a finite set of motion profiles, only a small number of representatives of Pareto optimal regions are chosen.

infinite control parameters (trajectories):
: For specified maneuvers, time-dependent control trajectories are required that, for instance, guarantee braking that fulfills prescribed conditions (e.g. braking time and distance). These curves have to be approximated by a finite number of continuous parameters in order to apply numerical optimization methods.

The (continuous time) dynamical behavior of mechatronic systems is typically modeled by ordinary differential equations of the form $\dot{\mathbf{x}}(t) = \mathbf{f}(\mathbf{x}(t), \mathbf{u}(t))$. Here, $\mathbf{x}(t)$ denotes the physical state of the system and $\mathbf{u}(t)$ the time-dependent control inputs, such as the braking forces. The function $\mathbf{f}$ is comprised of the physical principles drawn on in the technical system's model. These physical laws in turn depend on different kinds of parameters (constant, discrete or continuous). We assume that the continuous parameters can be quantized into regions for profile generation. In a real application, due to model discrepancies, optimal profiles would be combined with an underlying time-continuous controller. Thus, moderate additional deviations resulting from the quantization could be compensated for by this controller.

For convoy braking strategies, the vehicle mass can be examined to illustrate an example of a continuous parameter. In passenger as well as in freight RailCabs, mass can vary widely and does so in a continuous fashion. Each RailCab possesses an underlying lateral controller, which is crucial for traveling in convoys. Among others, this requires a large safety margin between the RailCabs. The controller ensures that the system maintains the optimal feedforward trajectory as long as the current RailCab mass deviates only slightly from the parameter value assumed in the model. An additional, smaller, safety margin in the compatibility function takes into account unavoidable deviations.

The feedforward optimal trajectories are solutions to optimal control problems. Such an optimal control problem is formally stated as

$$\min_{\mathbf{u}(t),T} J(\mathbf{u}(t),T) = \int_0^T C(\mathbf{u}(t))dt \tag{3.24}$$

with respect to

$$\dot{\mathbf{x}}(t) = \mathbf{f}(\mathbf{x}(t),\mathbf{u}(t)),\ \mathbf{x}(0) = \mathbf{x}_0,\ \mathbf{x}(T) = \mathbf{x}_T \text{ and } 0 \geq \mathbf{g}(\mathbf{x}(t),\mathbf{u}(t))\,\forall t \in [0,T]. \tag{3.25}$$

$J(\mathbf{u}(t),T)$ is the objective that has to be minimized and depends both on the control trajectory and on the final time of the control maneuver. The equations of motion, i.e. the ordinary differential equations, appear as constraints on the optimization problem. Boundary constraints are given by initial and final states $\mathbf{x}_0$ and $\mathbf{x}_T$, for instance the convoy speed at the beginning of the braking maneuver and the predefined braking distance at its conclusion. Additional constraints, e.g. bounds on the braking forces, can be modeled in the function $\mathbf{g}$.

Self-optimizing systems possess several optimization objectives that may become relevant during operation. This leads to multiobjective optimization problems (cf. *D.M.f.I.T.S*, [55], Sect. 1.2.3) which result in Pareto sets of optimal compromises between the concurring objectives. Therefore, our method includes the computation of a knowledge base of several Pareto optimal solutions corresponding to varying prioritizations of the objectives. We consider the objectives "optimize passenger comfort" (by minimizing acceleration) and "minimize braking time".

In real applications, optimal control problems have to be solved numerically. One class of numerical methods is based on direct discretization of the optimal control problem, by which, the original problem is transformed into a nonlinear restricted optimization problem that can be addressed by state-of-the-art software. Discretizing the optimal control problem means approximating the control trajectory with a relatively small number of continuous parameters. By using numerical integration schemes for the model simulations, the system's states can be discretized as well.

For the example application, optimal control trajectories for different RailCabs are computed. All of them are restricted by maximum control forces ($40\,kN$ for a passenger RailCab and $80\,kN$ for a freight RailCab), a convoy speed of $30\,m/s$ and a desired braking distance of $350\,m$. In Fig. 3.42, examples of braking trajectories for a freight RailCab are shown. Each control trajectory is defined by five points (equidistant in time) that are turned into a continuous signal via linear interpolation. This defines the input given to the system's dynamical model, by which the corresponding optimal state trajectories are generated.

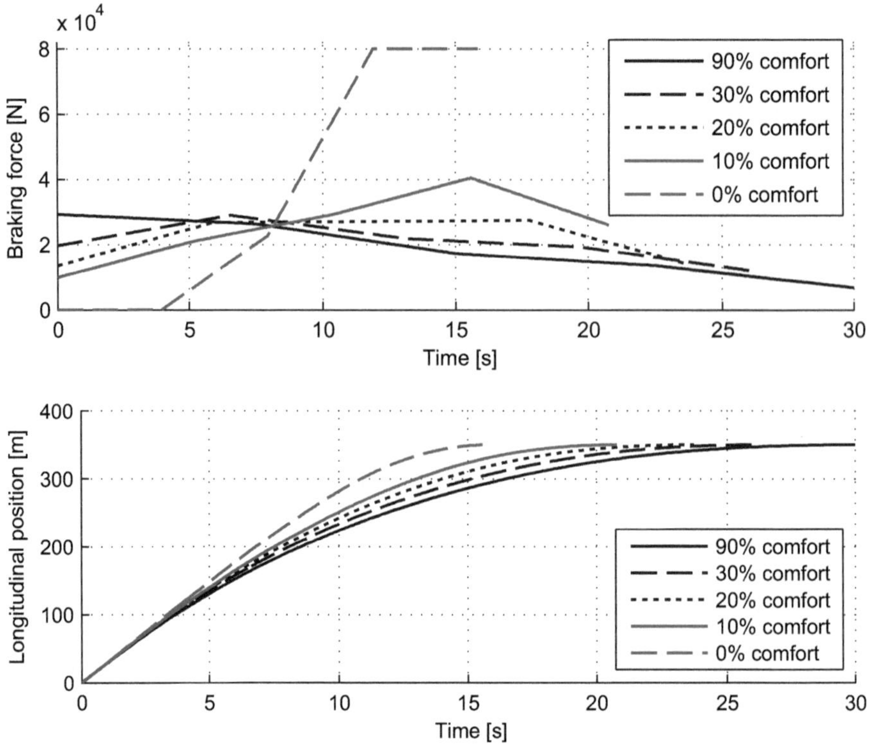

Fig. 3.42 Optimal control trajectories depending on an objective's priority ranking (top) and the resulting position trajectories (bottom)

Profile Compatibility

When a RailCab is traveling alone, i.e. not in a convoy, it can freely choose among the available motion profiles for each maneuver. This is because any condition that could invalidate a motion profile in such a situation would have been noted at design time and thus would have been removed (or rather, not been constructed in the first place by the optimal control algorithm). Therefore, the only safety-relevant restrictions that apply to the choice of a motion profile at runtime derive from the *interaction* of motion profiles from different RailCabs.

The most obvious example of such a situation is convoy travel. Each RailCab must choose motion profiles that fulfill certain *compatibility criteria* with respect to the motion profiles of RailCabs it will directly interact with, i.e. the RailCabs preceding and following in the convoy. One of the central ideas of this method is to encapsulate these compatibility criteria in simple algorithms (the *analytical functions*) to compare two motion profiles to each other.

These analytical functions have to return a boolean value, i.e. two profiles are either compatible or they are not. Furthermore, they have to be transitive, meaning that, if profile f is deemed compatible with profile g, and g is deemed compatible with profile h, then f must also be compatible with h. The analytical functions do not, however, have to be symmetric. Thus, even if f is compatible with g, g does not necessarily have to be compatible with f. This enables us to take the ordering of RailCabs on the tracks into account.

For each maneuver (e.g. controlled braking) and each situation (e.g. convoy travel), such an algorithm must be created. For controlled braking of a RailCab in a convoy, the required algorithm is fairly simple: assuming the minimum distance required for convoy travel, and taking into account the dimensions of each RailCab, do the adjacent RailCabs get closer to each other than a given minimum distance? This algorithm can easily be implemented by slightly modifying each motion profile to compensate for distance and form, subtracting the results and seeing if they ever drop below the safety margin. If so, the motion profiles are deemed incompatible; if not, they are compatible. This process is shown in Fig. 3.43.

Consider again our example of controlled braking maneuvers. It should be clear that if the motion profile of the RailCab joining the convoy is compatible with the motion profiles of both the preceding and following RailCabs, then adding the RailCab at that position will not make the controlled braking maneuver unsafe.

Thus, the question of whether the convoy will be able to safely execute maneuvers after the additional RailCab is added at the specified position is reduced to the question of whether the motion profiles are distributed in such a way that consecutive profiles for the same maneuver are always compatible.

Profile Distribution

In this method, correct distribution of motion profiles is ensured by communication protocols modeling message exchange between RailCabs. These communication protocols are executed before a RailCab enters a new situation, such as a convoy. First, the RailCab wishing to enter a convoy sends the set of its profiles to the convoy coordinator. Secondly, the convoy coordinator searches for a position where the RailCab may enter the convoy and which profiles it needs to use when doing so.

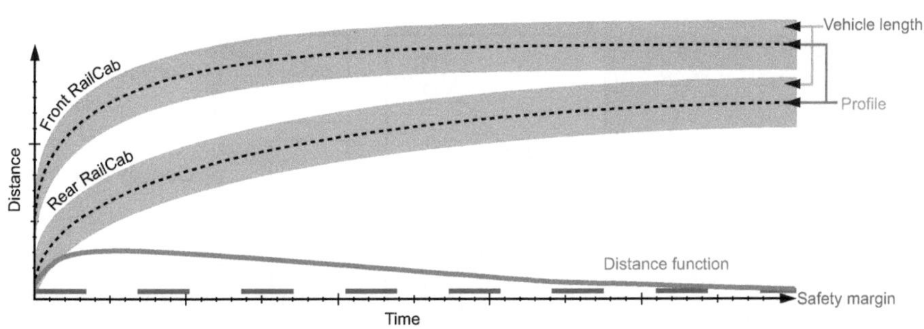

Fig. 3.43 Comparing two motion profiles with regard to minimum distance

We specify the communication protocol using real-time coordination protocols of MECHATRONICUML (*D.M.f.I.T.S*, [55], Sect. 5.2, [42]). Real-time coordination protocols use a state-based behavior specification technique called real-time statecharts to define this message exchange. Real-time statecharts are a combination of UML state machines [110] and timed automata [6] that enables the verification of real-time coordination protocols for safety and liveness properties.

In our example scenario, we have modeled the message exchange between a single RailCab attempting to join a convoy and a convoy of up to five other RailCabs. The modeling guidelines of MECHATRONICUML impose the restriction that the convoy coordinator is connected to all convoy members, but the convoy members are not connected to each other. This restriction enforces a kind of star topology (see Fig. 3.44). Fig. 3.45 shows the real-time statechart of the convoy coordinator. At runtime, the convoy coordinator has one thread executing the behavior defined by the real-time statechart *adaptation*. The convoy coordinator has one additional thread for each convoy member; this thread executes the behavior defined by the real-time statechart *sub-role*.

The manner in which the protocol implemented by these real-time statecharts works is loosely explained in the following. RailCab (*A*) wants to join a convoy led by RailCab *B*; it contacts *B* and proposes a merging maneuver using the message *requestConvoyEntry*. *B* either rejects the proposal outright (this may happen due to economic or safety reasons) by sending *declineConvoyEntry* or otherwise requests the set of all profiles that *A* possesses for all relevant maneuvers (using among others the messages *startProfileTransmission* and *profile*). *A* sends this information. Then, *B* invokes the operation *calculateProfiles* (circled in Fig. 3.45), which iterates all possible entrance positions for *A*. For each position, *B* goes through all maneuvers and checks whether *A* possesses a profile for every maneuver that is compatible with the profiles of the adjacent RailCabs for the corresponding maneuvers using the analytical functions. If possible entrance positions exist, *B* will use a heuristic to select one of them. The selected position is stored in the variable *newRailCabPosition* while the selected profiles are written to the variable *currentProfiles*. *B* sends the position and the selected profiles to *A* using the message *enterConvoyAt*; *A* can then

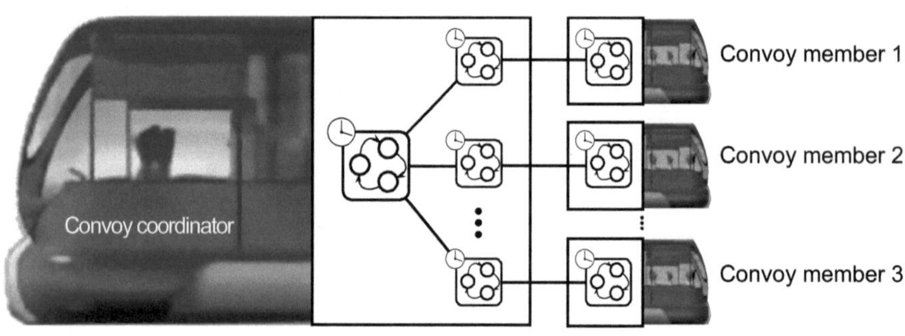

Fig. 3.44 The communication structure in the Convoy Merging example

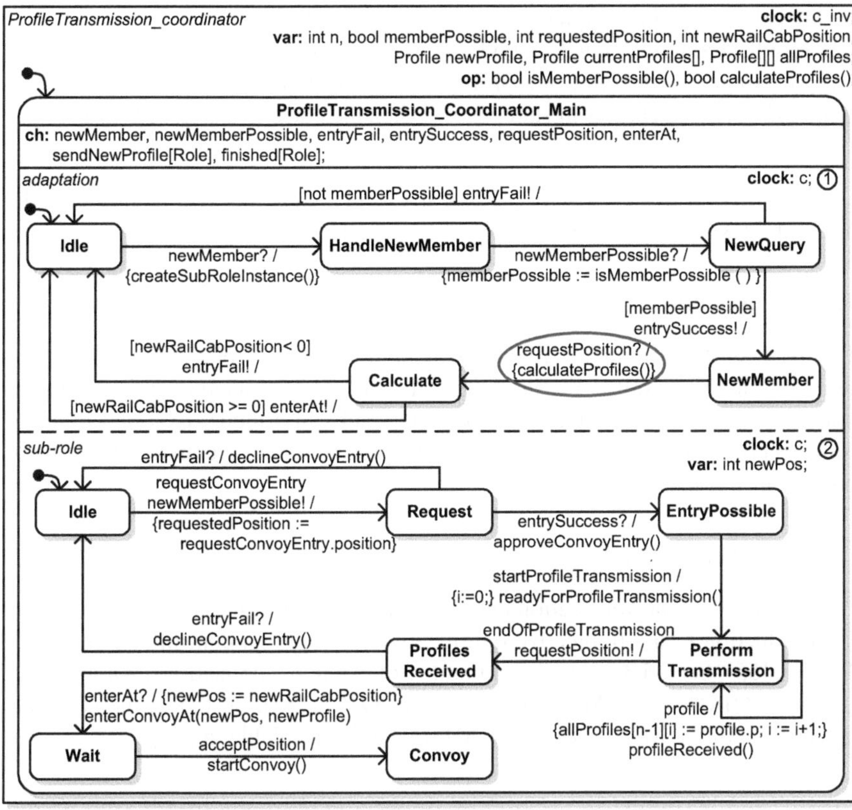

Fig. 3.45 Real-time statechart of the convoy coordinator

initiate the merging maneuver. If no locations have been deemed safe, *A*'s proposal is rejected and the process ends.

The operation *calculateProfiles* of the real-time coordination protocol calls the implementation of the analytical functions as an external method. The verification of the real-time coordination protocol thus guarantees the following: If the analytical functions are correctly implemented, i.e. they return true if and only if the given profile combination is safe from a mathematical perspective, then the distribution of motion profiles across a convoy performed by this real-time coordination protocol guarantees safe operation of the convoy as a whole.

The MECHATRONICUML model of the communication protocol can be used as an input for formal verification using, for example, the design-time verification introduced in *D.M.f.I.T.S*, [55], Sect. 5.2.2, or the timed model checker UPPAAL [16]. We have verified the model for a convoy of 5 RailCabs; we have also successfully formalized and verified the following properties given in natural language using UPPAAL.

- The real-time coordination protocol is free of deadlocks.
- If a RailCab sends *requestConvoyEntry* to the coordinator, it receives a *startConvoy* message or a *declineConvoyEntry* message eventually.
- If a *sub-role* statechart is in state *convoy*, then the corresponding convoy member is also in state *convoy*.

The restriction to a convoy of 5 RailCabs is a result of limitations in the model checker UPPAAL. It is not imposed by the method itself.

3.2.10.4 Results

The first result of this method is a set of motion profiles for every maneuver in the input; another result is the implementation of a set of analytical functions for comparing said profiles. Finally, the method results in a model of a real-time coordination protocol designed to distribute the profiles across a convoy according to compatibility constraints.

3.2.10.5 Further Reading

The application of the verification method with motion profiles is described in [52]. An introduction to multiobjective optimization and optimal control is given in *D. M.f.I.T.S*, [55], Sect. 5.3, including a number of example applications. An introduction to MECHATRONICUML is given in *D.M.f.I.T.S*, [55], Sect. 5.2; the complete specification of the MECHATRONICUML method is available in [15].

3.2.11 Dependability-Oriented Multiobjective Optimization

Peter Reinold, Walter Sextro, Christoph Sondermann-Woelke, and Ansgar Traechtler

The goal of this method is to evaluate the dependability of a self-optimizing system by considering one or several dependability-oriented objectives in a multiobjective optimization process. The Multi-Level Dependability Concept (M-LD Concept) described in Sect. 3.1.2 is used to define the necessary weighting of dependability-oriented objectives. With this concept, it is possible to weight or prioritize objectives related to the dependability attributes reliability, availability and safety, depending on the momentary situation.

This section describes a method of strengthening the Design and Development of the M-LD Concept; the method described analyzes the effect(s) that dependability-oriented objectives, derived from multiobjective optimization, can have on the dependability of the system. The results of the analysis are used to refine the integration of the aforementioned objectives into the M-LD Concept.

Additionally, the effects of possible failures of sensors and actuators are considered. In particular, failures of actuators can be integrated in the optimization process as constraints; however, this aspect is only relevant for redundantly actuated systems which can compensate for such failures.

3.2.11.1 Prerequisites and Input

This method is based on the method *Early Design of the Multi-Level Dependability Concept* (Sect. 3.1.2), which identifies dependability-oriented objectives based on the partial model *System of Objectives* that should be considered within the M-LD Concept. To take full advantage of this method, a redundantly actuated system and/or a system with sensor redundancies is needed. These necessary redundancies are often already installed in safety-critical systems to compensate for the failures of individual components. Further information about the sensors and actuators used can be found in the partial model *Active Structure*. Information concerning the event and the type of a failure is additionally required. The method *Early Probabilistic Reliability Analysis of an Advanced Mechatronic Systems based on its Principle Solution* (Sect. 3.1.1) assists in determining possible failures. It is assumed that actuator or sensor failures can, in fact, be detected and that if one such failure occurs, the particular actuator or sensor in question nonetheless has a defined behavior, i.e. it is not performing an unknown movement. A suitable model including information about the system and the reliability of the different components themselves (especially the probability of failure for each relevant component) is also needed. This model can be derived from the partial models *Environment*, *Application Scenarios*, *Active Structure* and *Shape*.

3.2.11.2 Description

The method consists of two principle areas, as described in the following. The first deals with multiobjective optimization and focuses on dependability-oriented objectives and the integration of actuator failures as additional constraints within this optimization. The second consists of a reliability assessment of redundant system structures and uses the M-LD Concept to assess different system states.

For the first part, the dependability-oriented objectives are identified in the partial model *System of Objectives* and are included in the model of the system dynamics. In addition, possible actuator failures are integrated as constraints in the multiobjective optimization. The modeling of a failing actuator is reflected in the optimization, for example by constraining the force of this actuator to zero for that case. The Pareto sets are calculated using model-based multiobjective optimization (cf. Sect. 1.1.2.1). Using the optimization results, the possibilities of altering the system behavior in order to improve dependability are analyzed.

The reliability assessment in the second part of this method focuses on the redundant structures in the system, beginning with a qualitative analysis in which the different combinations of operable and inoperable components are identified. These combinations are evaluated and afterwards classified according to the M-LD Concept. In Level I of the M-LD Concept, the system is functioning as desired and is failure-free. The weighting of the objectives can be chosen without special consideration of the dependability. At Level II, one or more redundant components are already non-functional. Thus, the weighting of dependability-oriented objectives is increased. In Level III, the situation becomes more critical; safety becomes the most important aspect and the functionality of the system may be reduced to avoid additional failures and dangerous situations. Level IV is defined by the fact that the system is in danger of becoming uncontrollable should an additional failure occur. To avoid this situation, the system is forced into a fail-safe-state upon reaching Level IV.

Two types of failures can be differentiated: On one hand, there are failures for which the rate of failure increases over time. These failures are primarily caused by wear and tear; prognostic models are used to predict the remaining useful life of the component. On the other hand, there are failures with a constant failure probability, such as cable breaks, etc. These failures are more or less unpredictable, but often require a rapid reaction, e.g. adaption of the system. Both types of failure are considered in a statistical evaluation to calculate the failure probability of the entire system. This reliability assessment aids in classifying the momentary configuration as one of the four levels of the M-LD Concept. This classification influences the selection of priority levels for system objectives within the self-optimization process. It is implemented during design, but is also used during operation to assess the situation. In such conditions, the assessment considers existing failures to compute the current failure probability.

3.2.11.3 Results

The results can be summarized by two important considerations. For one, combinations of failures are assessed by reliability methods and afterwards classified by the M-LD Concept. These results are used during operation to evaluate the current system status. In addition to this, a system of several objectives is also analyzed and the influence of each dependability-oriented objective is simulated. As part of the multi-objective optimization process, actuator failures in particular are taken into account by including them as constraints in the optimization. Thus, a detailed analysis of the behavior of the system in case of sensor or actuator failures can be provided. This method enables the system to react to failures of redundant components adequately and thus, by implementing counter-measures, to improve its dependability.

3.2.11.4 Application Example

The method explained above has been used to increase the dependability of the test vehicle "Chameleon" (cf. Sect. 1.3). Only the test vehicle's horizontal dynamics are considered. Due to its special actuator concept, the "Chameleon" can decelerate in several ways, by turning the wheels inwards, by using the driving motors as generators, or by some combination of the two. With these possibilities it is up to the information processing unit to decide which possibility is optimal for the current situation while also considering possible actuator failures and wear. In this application, the effect different prioritizations of the objectives have on the reliability of the system is of particular interest. Another focus here is on the use of existing redundancies at runtime. Depending on the curent situation, suitable objective functions (e.g. reducing tire wear) and the appropriate weighting of objectives is selected.

This guarantees that the necessary forces can be optimally distributed among the four wheels even in changing driving situations.

Any desired maneuver is described by the yaw rate, the velocity and the slip angle. By using the inverse dynamics, it is possible to compute the necessary longitudinal and lateral force as well as the desired yaw moment to effect the maneuver. Based on the model, there are eight tire forces (longitudinal and lateral force of each tire) to produce the desired forces in the center of gravity. Hence, there are redundant possibilities of completing the driving task, i.e. there are degrees of freedom for the allocation of the tire forces. Within the technical restrictions, it is possible to use these degrees of freedom to optimally produce the desired movement. Despite this freedom, the system is subject here to the constraint that the execution of the desired movement must be guaranteed. Optimization objectives are the minimization of the tire wear, the energy consumption and the most effective utilization of the limited transmittable forces due to friction. In the following, a braking maneuver while traveling straight forward is analyzed. As described above, there are several possibilities for braking the vehicle. For decelerating while moving forward, there are no conflicts between the optimization objectives: for all of the above-mentioned objectives, it is optimal to brake by using the driving motors as generators. However, in order to achieve stronger deceleration, it is necessary to turn the wheels inwards in addition to reversing the motors..

The vehicle has four steering motors and four driving motors. Each of them can be functional or non-functional, giving a total of $2^{(4+4)} = 256$ different actuator configurations. The failure probability of the whole system is computed using a Bayesian network. Similar to a fault tree, the top event "system failure" is composed of logical connections between the components. The failure probability of each component is calculated for the current point in time and the system probability is calculated by the predefined logical connections; in addition, combinations of actuator failures are assessed. Based on the reliability assessment of the system, the multi-dependability concept is used to assess the system status in case of actuator failures. As a basis for this, the reliability assessment conducted across the Bayesian networks is used in the operating phase as well to assess the system status. Through the dependability-oriented optimization, actuator failures can be counterbalanced. The M-LD Concept is used to derive suitable measures based on the assessment of the current system status. In Level I, all actuators are functional and, in most cases, the focus is not on the dependability-oriented objectives. In Level II, one or more actuators has failed. The objectives of minimizing the actuator loads are taken into account in addition to the normal objectives. If there are only two possibilities remaining for decelerating the vehicle, Level III is reached and additional measures, such as protecting the critical actuators by reducing the speed of the vehicle, are considered. Level IV is reached if only one possibility of decelerating the vehicle is still functional. In this case, the vehicle is forced to a fail-safe-state, meaning it will be brought to a stop.

The vehicle is intended to perform a desired maneuver in an optimal way; during operation, the control strategy can increase safety by using the redundancy provided by the vehicle: actuator failures can be included as additional constraints. This idea can be illustrated by an example: the driving motor of the front wheel on the left-hand side may be out of action. The constraint connected to this failure is that the driving force of the left front wheel is zero. By accepting this as a constraint, the optimization result will guarantee the desired movement of the vehicle. Fig. 3.46 presents the simulation results for braking in this situation. The applied braking force for the vehicle should be 200N. The vehicle is traveling in straight line while braking, so the values for lateral movement as well as the yaw rate are zero. If all actuators would be functional, the total braking force of 200N could be allocated to the wheels equally, i.e. each wheel would provide a braking force of 50N. The figure presents two possible options for a braking maneuver in the case of the supposed failure. For each wheel and each option, the braking force produced by the driving motor F_{dm} and those produced by the steering motors F_{sm} are given. In the first option, braking is achieved by the driving motors alone. To compensate for the actuator failure of the left front wheel, the left rear wheel has to brake with the doubled amount of longitudinal force than in the case of a fully functional vehicle, which in this case is 100N. The right front and rear wheel brake with 50N each to achieve the desired braking force of 200N. The second option uses the symmetric inward turn of all four wheels. This option could make sense if further constraints are imposed on the driving motors. The force produced by the driving motors is significantly reduced compared to the first option. Online, the M-LD Concept chooses the suitable option.

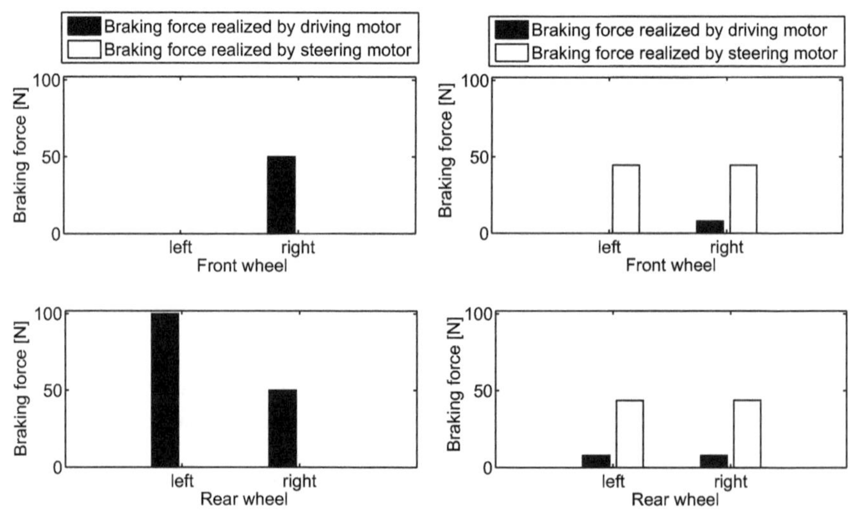

(a) Exemplified Option 1 (b) Exemplified Option 2

Fig. 3.46 Compensation for the failure of the left front wheel driving motor

3.2.11.5 Further Reading

The method and this example application are also described in [141]. A possible maneuver accounting for the breakdown of a steering motor and the suitable optimization is explained in [128]. The focus on dependability-oriented system objectives is also described for different applications. In [101] the objectives and behavior in the case of actuator failures for an active suspension module are analyzed (see also Sect. 3.2.1); sensor failures are discussed in [140].

3.2.12 Self-healing in Operating Systems

Katharina Stahl

In the context of self-optimizing systems, the operating system has to cope with complex and changing behavior of dynamically reconfiguring hardware and software. As part of its functionality, it has to ensure the reliability of the entire system.

Methods such as Hazard Analysis (described in Sect. 3.2.8) and Online Model-Checking (described in Sect. 3.2.14) aim to verify and control the correct execution of the software system. However, these methods basically rely on specification knowledge or an system model generated offline. The system state analysis is computed based on an exhaustive system model, and is therefore resource-intensive. For this reason, it can only be executed offline or outside of the system while the system continues to operate.

Furthermore, the behavior of systems that operate in a self-organizing manner in uncertain environments can lead to unforeseen and unpredictable system states. Autonomous reconfigurations of the system may produce system states that lead to unstable or malicious system behavior. Self-organization opens up some degree of freedom for system behavior, so that system designers and development tools are not able to determine all potential system states in the specification model. This leaves a gap in terms of dependability at runtime.

To ensure dependability of the entire system, the operating system requires mechanisms to cope with dynamic behavior and to monitor system behavior at runtime in order to identify potentially malicious system states caused by autonomous reconfiguration, and potentially being unidentified by any specification or model. With our self-healing framework, the operating system is able to profile system behavior. It builds up a knowledge base of normal system behavior at runtime without predefined system model. This knowledge base is incorporated directly into the operating system, and thus is continuously updated. The framework provides a mechanism for detecting deviations from normal system behavior. In this context, the self-reconfiguration abilities of the applications and the hardware constitute an additional challenge, as in a self-reconfigurable system it is hard to distinguish between intended and malicious deviations from normal system behavior. Further properties and information about the system state are considered for the evaluation of the anomalous system behavior in order to identify if it could potentially degrade the system's dependability. The main objective of the framework is therefore to maintain, or at least re-establish the system's overall performance and service delivery by reconfiguring the operating system.

3.2.12.1 Prerequisites and Input

ORCOS (Organic ReConfigurable Operating System [48]) is a highly customizable and (re-)configurable real-time operating system (OS) for embedded systems. We use this platform to implement self-healing properties in the operating system.

Customizability of the Operating System

The operating system ORCOS is composed of kernel modules that can be configured individually at compile time. For example, the scheduling strategy, the memory management method, or even hardware-dependent functions can all be set up according to system requirements by an XML-based configuration language SCL (Skeleton Customization Language, see [48]). Dependencies between particular kernel modules can be specified to ensure correct functionality and dependability of the customized operating system. In resource-restricted environments such as mechatronic systems, efficiency is an important factor. Configurability ensures that only the necessary code will be compiled into the executable and loaded onto the device.

3.2.12.2 Description

The basis for integrating such a self-healing framework into the operating system (OS) is an adequate OS architecture that encompasses both the ability to monitor and analyze system behavior, and the ability to reconfigure the system in order to react to a malicious system state. This section first describes the architecture of the real-time operating system and then the according components that constitute the self-healing framework.

The main challenge of this approach is the problem of anomaly detection. Within this context, we define what actually determines system behavior and which internal and external factors have an impact on and can cause changes in the system's behavior. Then, we present our approach for online anomaly detection within the self-reconfiguring system, which relies on the Danger theory [99] of Artificial Immune Systems (AIS) [27,31]. The applied algorithm and further adjustments on the OS architecture that are necessary to implement the AIS algorithm are presented in the later part of this section.

Operating System Architecture

To integrate self-healing capabilities into the operating system ORCOS, we have adjusted the OS architecture following the example of the Observer-Controller Architecture first instantiated by the Organic Computing Initiative [130].

We have expanded the ORCOS architecture to include an observer component that is responsible for monitoring the system and for collecting and providing knowledge about the system behavior.

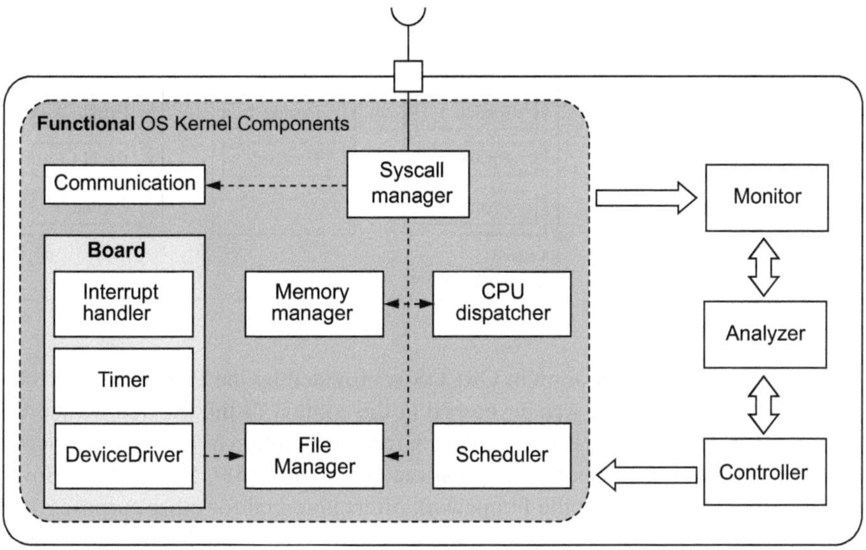

Fig. 3.47 Architecture of a self-healing operating system

Although the Observer implements two functions that are strongly coupled (collecting data and analazing), they exhibit self-contained tasks which can be executed in a timely decoupled manner. Therefore, the Observer is subdivided into the two separate entities: a Monitor and an Analyzer. The Monitor collects behavioral data and aggregates the data for the Analyzer. Then, the Analyzer evaluates the data and passes its evaluation results to the Controller. Based on the results of the analysis, the Controller is responsible for planning and executing decisions for system reconfiguration.

The resulting architecture of the operating system ORCOS is shown in Fig. 3.47. In this operating system design, the self-healing framework is strongly isolated from the remaining functional OS kernel components. This isolation ensures that the functional OS kernel components are not integrated into the self-healing process, so that for the functional OS kernel components the execution of the self-healing framework can take place in an imperceptive manner. Furthermore, the self-healing framework must not degrade the execution of the real-time applications in terms of reaching their deadlines so that it can only be implemented having soft real-time requirements or at lowest system priority.

Reconfiguration

In the context of dependability and self-healing, reconfiguration is used to reestablish an acceptable system state where, for instance, components exhibiting faulty behavior can be replaced by other components. Hence, reconfiguration essentially requires the existence of alternatives for the Controller.

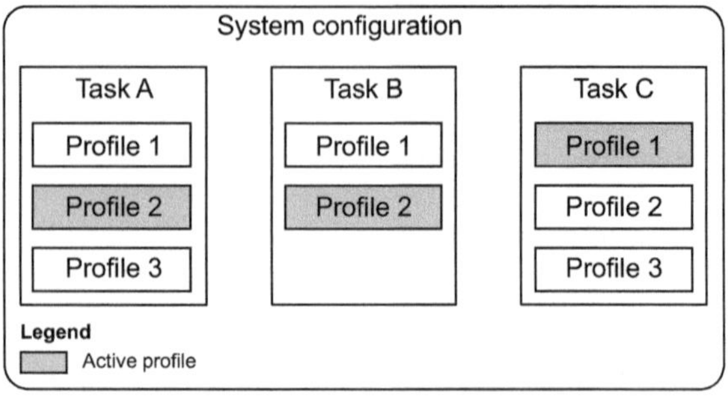

Fig. 3.48 Profile Framework: example of a system configuration

The basis for reconfiguration in ORCOS is provided by the Profile Framework. Originally, this framework was developed in the context of the Flexible Resource Manager (FRM) [109] to self-optimize resource consumption in resource-restricted real-time systems (for further description see *D.M.f.I.T.S*, [55], Sect. 5.5.2). However, the principle of the Profile Framework offers potential for other purposes: the Profile Framework enables alternative implementations of an application task. Each alternative implementation is represented by a defined profile for that task. Implementations may vary, as originally intended in the FRM, by different resource requirements.

At each point in time only one profile of a task is active (see Fig. 3.48). A configuration c of the system is defined as Configuration $c = (p_1, p_2, \ldots, p_n)$, with n being the number of running tasks τ, $p_1 \in P_1, p_2 \in P_2, \ldots, p_n \in P_n$ and P_i being the profile set of task τ_i. Each task must define at least one profile to be executed.

For the purpose of our self-healing framework, we can extend this definition: profiles are not only restricted to tasks, but are also defined for OS components.

The former definition of profiles is extended so that profiles may now differ in their demand for resources. in the choice of which resources are applied (e.g. a specific communication resource), in the implemented algorithm (e.g. in terms of accuracy of the algorithm or strategy), in execution times, in deadlines etc.

During runtime, the Controller may switch between the profiles of a task or a OS component in order to reconfigure the system according to some system restrictions. A reconfiguration is required when the analyzing algorithm detects anomalous system behavior, which, for example, could be caused by a defective hardware resource and could lead to malicious behavior. In such a case, the Controller must deactivate all profiles that use that resource and, consequently, it must activate another profile for each affected task or OS Component. To maintain flexibility of the operating system at runtime, an extension of the operating system kernel allows executable OS modules to be uploaded online and thereby to add new profiles for this component. Kernel components (e.g. the memory manager) can be exchanged based on changing requirements of the self-optimizing system.

System Behavior

The state of an operating system consists of a range of system status information, such as the CPU usage, resource and memory consumption, values of kernel parameters, etc. These system state parameters are discrete values for a specific time stamp. However, behavior is usually understood as a course of actions and derived from an observable course of changes within discrete system states.

The operating system is a service platform for applications that are running on the system. Hence, the characteristics of these applications have a major impact on the operating system state parameters. Due to security aspects, ORCOS provides system calls as the only interface for the applications to the operating system so that the system state and the derived system behavior are strongly dependent on the applications' system calls.

Based on this, we define that the operating system behavior is determined by (sequences of) system call invocations executed by the applications tasks with the associated information (arguments, return values, return addresses etc.) .

External Impact on System Behavior

Autonomous reconfiguration of the operating system is not the only challenge when considering the entire system. Self-optimization is present in all system layers. Therefore, our operating system must additionally deal with self-reconfiguring hardware and self-reconfiguring software. A reconfiguration, either in the underlying hardware or in the overlying software layer, will obviously cause deviations in the normal system behavior. Even if the particular reconfiguration was not induced by the operating system itself, it was intended by system-internal source, meaning that the behavioral changes must be accepted as non-malicious. Therefore, interfaces must exist that allow other system layers (such as the hardware or software) to inform the OS whenever a reconfiguration has been initiated by a source outside its own control (by e.g. interrupts or signals).

Inspiring Paradigm

The self-healing mechanism depends heavily on the efficiency of the algorithm used for runtime analysis and behavior evaluation. As we have to deal with dynamically changing system behavior in a resource-restricted environment, a suitable approach must possess certain qualities, such as:

- disposability,
- autonomy,
- dynamical adaptivity,
- abililty to cope with new system states,
- learning mechanism for remembering correct and malicious states, and
- simple algorithms with a low resource consumption.

Considering the context of self-healing systems, Artificial Immune Systems (AIS) [27, 31] are a good source of inspiration for problem solving.

Self-healing Framework

The workflow of the self-healing framework in the operating system is defined by:

Monitor:

Step 1:
Data collection of system state and behavior data, composed of :
- data about OS parameters: CPU utilization, resource usage, etc.
- data about tasks being executed: system calls, system call arguments, return addresses and other related information

Step 2:
Data interpretation and generation of a system behavior representation

Analyzer:

Step 3:
Identification of system behavior; detection of deviations/anomalies by pattern matching

Controller:

Step 4:
Analysis of results from behavior identification and evaluation of effect on the overall system

Step 5:
Reaction to system behavior, in particular to behavioral anomalies: reconfiguration of the system

Danger Theory from Artificial Immune Systems [2, 99] offers appropriate features to serve as an inspiration for implementing the self-healing framework in our dynamically reconfigurable system. Dendritic Cells build up the core of this population-based approach. In immunology, a Dendritic Cell (DC) initially serves as an Antigen Presenting Cell (APC), which means that this cell signals the presence of substances that indicate an anomaly and stimulates an immune response.

During its lifetime, a DC can obtain three different states: *immature*, *mature* and *semi-mature*. The initial state of the DC is *immature*. Residing in this state for a particular period of time, the DC will observe and examine the structures for which it is specified. Then, the DC migrates either into the *semi-mature* or into the *mature* state. This migration is illustrated in Fig. 3.49. The decision for state migration depends on two factors: first, its own evaluation of the observation, and second, input signals from the surrounding system.

According to Danger Theory, the following input signals are defined:

Safe signal:
indicates that no threat has been identified in the system

PAMP (pathogen-associated molecular pattern) signal:
indicator that a known threat has been localized

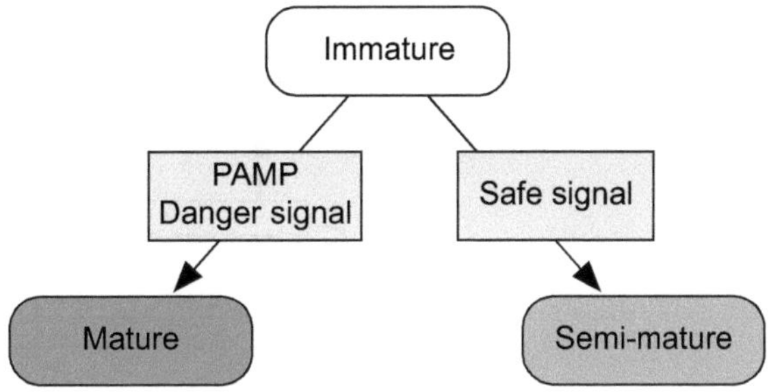

Fig. 3.49 DC state migration in correlation to input signals

Danger signal:
indicates a potential danger suspected due to local behavior deviations

Inflammation signal:
general alarm signal

The DC'S state transitions are probabilistic decisions that depend heavily on several thresholds. To decide on a state transition, the DC sends out an output signal reflecting its local evaluation and adjusts the system input signals.

We are deploying Dendritic Cells for self-healing in the operating system with the objective of evaluating the behavior of the system entity that the DC is assigned to monitor. In this context, we have defined the system behavior using OS state information and system call invocation by tasks. Hence, we use a DC for profiling the behavior of a task or a specific OS kernel property. Referring to the self-healing workflow described above, the set of DCs takes over the responsibility of the Monitor in a distributed manner.

Local evaluations by DCs are collected by the shared DCA Monitor. Additionally, the DCA Monitor supplies the DCs with values of input signals from a central location and stores the knowledge base containing the normal behavior profiles, as well as already-detected dangerous system states (PAMP); the latter are recorded in order to enhance immediate detection of known dangers. The resulting architecture for the self-healing framework in the operating system is illustrated in Fig. 3.50.

Each DC starts its local evaluation in the *immature* state. In this state, the DC samples selected system behavior data for the behavior analysis and evaluation. The process of data sampling of a DC is limited by a predefined data amount or by a timing condition. Parameters are monitored and collected in order to establish system behavior data based on the requirements of the analyzing algorithms. The data set is configurable at runtime according to the analyzing method. As any OS component of ORCOS is configurable and exchangeable at runtime, the analyzing algorithm can also be exchanged according to the given restrictions (e.g. available free resources for execution).

After data collection, the behavioral data has to be analyzed. Therefore, each DC provides a *normal behavior profile* for the component (e.g. task) it is monitoring. The objective of the data processing is to identify deviations in the real behavior from the *normal behavior profile*. The DC executes a (simple) pattern-matching mechanism in order to pre-evaluate the real local behavior monitored. Based on its

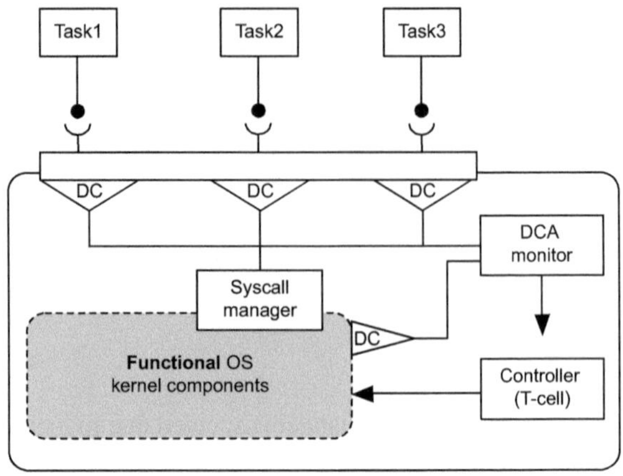

Fig. 3.50 Resulting architecture for a DC-based operating system

local knowledge, a DC classifies the behavior; if the real behavior complies with the *normal behavior profile*, the actual behavior is classified as safe. In that case, the DC switches to the *semi-mature* state and amplifies the value of the *safe signal*.

If the local behavior is outside the range of the *normal behavior profile*, in the first instance the system behavior is classified as suspicious. From here, the DC can either migrate into the *semi-mature* or the *mature* state. The migration of the DC requires further evaluation that is related to the system context as represented by the system's input signals.

If the local behavior corresponds to a system behavior that has been already identified as malicious or dangerous, then the DC migrates to the *mature* state and outputs a *PAMP* signal. If the current local behavior exceeds the range of the *normal behavior profile* and is combined with the presence of the *danger signal*, the DC tends to migrate to the *mature* state and consequently increases the value of the *danger signal*. If, on the other hand, the actual current behavior is outside the range of the *normal behavior profile* with no *danger signal*, but rather within a dominant presence of the *safe signal*, then the behavior tends to be tolerated. Consequently, the DC migrates to the *semi-mature* state.

In order to inform the OS about a reconfiguration at the software application or the hardware layer, the system provides the *inflammation signal* as a general alarm signal. The occurrence of the *inflammation signal* instructs the DCs to build up new system behavior knowledge which is based on the behavior produced by the new system configuration.

In the context of AIS, the Controller is an immune T-cell that is responsible for the immune response and, thus a reaction to the present system behavior. It receives the collected information from the DCA Monitor and decides - based on thresholds and predefined conditions - whether the observed behavior may lead to a system failure. As already described, it can then initiate a system reconfiguration, if necessary.

3.2.12.3 Results

The self-healing framework offers the operating system the ability to monitor and analyze system behavior. Its main potential is for evaluating unspecified and unknown system states for which there is no previously specified behavioral model. In correlation with environmental input signals, the framework ensures that behavior evaluation results are not determined based only on local information. Thus, this AIS-inspired approach substantially enhances the system's dependability using a runtime method with low computation efforts.

3.2.13 Self-healing via Dynamic Reconfiguration

Sebastian Korf and Mario Porrmann

The goal of this method is to increase the reliability of a self-optimizing system by taking advantage of the reconfigurability of the underlying hardware. Reconfigurable devices such as FPGAs are increasingly being used in several areas of application. With the advent of the new generation of SRAM-based partially reconfigurable FPGAs, the implementation of System on Programmable Chips (SoPCs) has become a feasible alternative to traditional options [83]. Nowadays, it is possible to implement a comprehensive system which embeds hardwired microprocessors, custom computational units, and general-purpose cores on a single chip. This technology is able to cope with today's requirements for a short time-to-market and high resource efficiency.

Since nanoelectronics with feature sizes of 45 nm and below has become reality for reconfigurable devices, designers have to consider the fact that faults – static and dynamic – will increasingly affect their products. Hence, yield and reliability are becoming key aspects in SoPC design [24]. These architectures are receiving increasing interest from various application domains. Even safety-critical missions, driven by avionics and space applications, are especially attracted to using SoPCs due to low non-recurring engineering costs, reconfigurability and the large number of logic resources available. One of the most significant problems in choosing to use SoPCs in safety-critical applications are the effects induced by different types of radiations such as alpha particles, atmospheric neutrons and heavy ions [118]. These particles may induce non-destructive loss of information within an integrated circuit, provoking Single Event Upsets (SEUs) [138].

Partial dynamic reconfiguration, i.e., changing parts of a reconfigurable fabric at runtime while leaving other regions untouched, can be used to further increase resource efficiency and flexibility of SoPCs [116]. Additionally, dynamic reconfiguration can be used to correct a corrupted configuration of a device, caused, for example, by an SEU, at runtime. Together with continuous monitoring of the configuration memory, this enables self-healing capabilities for the implemented information processing system. Furthermore, partial reconfiguration can be used to implement adaptive redundancy schemes. Based on sophisticated error monitoring and on user-defined security levels for each implemented hardware module, redundancy can be adapted at runtime. Critical modules can, for instance, be implemented with triple modular redundancy (TMR) [142]. Depending on changing environmental conditions and on the application requirements, the level of redundancy can be dynamically adapted for each module.

3.2.13.1 Prerequisites and Input

For the effective implementation of dynamically reconfigurable systems, we proposed a layer-based approach in *D.M.f.I.T.S*, [55], Sect. 5.4.3.1. This model systematically abstracts from the underlying reconfigurable hardware to the application level by means of six specified layers and well-defined interfaces between these layers. The main objective of this model is to reduce the error-proneness of the system design while increasing the reusability of existing system components.

Typically, a partially reconfigurable system is partitioned into a Static Region and a Partially Reconfigurable Region, abbreviated as PR Region (Fig. 3.51). The configuration of the Static Region is not changed at runtime; all static components of the system are located in the Static Region (e.g., the reconfiguration manager or the memory controller). The PR Region is used for runtime reconfiguration, and all dynamic system components are located in a PR Region; a partially reconfigurable system can be composed of one or several separate PR Regions.

In addition to the partitioning of the FPGA, the concept of partial reconfiguration requires a suitable communication infrastructure for connecting the PR modules and the Static Region. The communication infrastructure should not introduce any further heterogeneity in the system; this is so that the flexibility of placement is maintained by preserving the number of feasible positions of the PR modules. Homogeneity implies that the individually reconfigurable resources are connected via

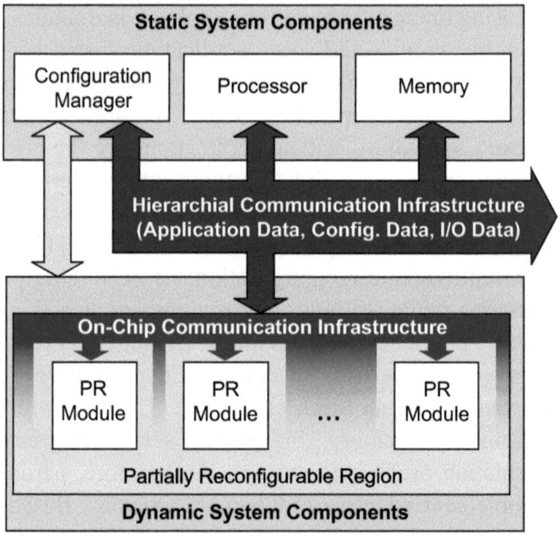

Fig. 3.51 Overview of a partially reconfigurable system divided into static and dynamic system components

the same routing infrastructure. Thus, modules cannot only be placed at one dedicated position, but at any position with sufficient free contiguous resources [66].

In addition to utilizing the reconfigurable resources, applications can also be mapped onto the embedded processor of the SoPC. The hierarchical communication infrastructure ensures that all system components can access the internal memory, which stores all relevant data for applications, IO, and PR module configuration.

3.2.13.2 Description

In our approach, dynamic reconfiguration is provided by a combination of dynamically reconfigurable hardware and a reconfigurable real-time operating system (RTOS), running on the embedded processor of the SoPC. While the proposed hardware platform offers the fundamental mechanisms that are required both to execute arbitrary software and to adapt the system to new requirements (e.g., by dynamic reconfiguration), the RTOS provides an interface between the hardware and the software application and decides whether a task will be executed in software, in hardware, or in a combination of both. Thus, depending on the actual environmental conditions, the high flexibility of the approach can be used to optimize the SoPC for energy efficiency or resource utilization.

Furthermore, the high flexibility of the approach can be utilized to increase the dependability of the system. This can be done on three different levels: system level, module level, or gate level. The selection and combination of appropriate methods depends on the desired level of reliability and the expected failure rate.

System Level:

A basic requirement for PR module placement is the availability of sufficient contiguous resources in the PR Region of the FPGA. In case of a detected failure of a complete FPGA, PR modules located on this FPGA can be migrated to other FPGAs. If the amount of available resources is not sufficient, modules that have low priority can be replaced by simpler modules with lower resource requirements, or can even be removed from the system entirely. The goal of this approach is to provide sufficient resources for high-priority modules so that a basic functionality of the system can be guaranteed. Additionally, spare FPGA devices can be integrated into the system to further increase reliability.

SEUs are especially critical for reconfigurable devices, since they can affect the configuration memory and therefore change the behavior of the system. A frequently occurring reconfiguration (blind scrubbing) can increase the reliability, as newly configured SRAM cells are correctly set and potential bit-flips are removed. In addition to blind scrubbing, a continuous readback of the configuration bitstream combined with an online integrity check can be implemented (readback scrubbing). While blind scrubbing does not provide any information about detected faults, readback scrubbing enables continuous monitoring of the failure conditions.

Module Level:

Dynamic reconfiguration enables the implementation of different, adaptive levels of redundancy and reliability on the level of PR modules. Components in which faults are unacceptable could be implemented using very safe but space-consuming techniques such as TMR. In less critical PR modules, faults may be acceptable, as long as they can be detected and corrected within a given time frame. Furthermore, scrubbing, as described above, can be performed on module level. Critical PR modules can be checked or reconfigured more often than less critical PR modules.

Gate Level:

Also permanent faults can be detected and located using the methods described above. Once a permanent fault has been located, it can be considered as non-usable during module placement. Therefore, the complete device does not have to be turned off, but rather, only a single module is deactivated. This module can even be used for future PR modules, if the defect is taken into account (i.e., masked out) during PR module implementations.

3.2.13.3 Results

Partial dynamic reconfiguration can be used to increase the flexibility and resource efficiency of microelectronic systems. The proposed strategies for scrubbing, module relocation, and adaptive redundancy enhance the reliability of the whole system at much lower cost than traditional approaches, such as triple modular redundancy of the complete system architecture.

3.2.13.4 Application Example

Methods of increasing the reliability of microelectronic systems will be of increasing importance when utilizing nanoscale semiconductor technologies in the future. However, even today, special system architectures are necessary if high dependability is a major concern (as in automotive and aerospace applications) or when targeting operation in harsh environments. Information processing in space combines the aforementioned requirements and has therefore been chosen as an example of using dynamic hardware reconfiguration for self-healing.

Performance requirements for onboard processing of satellite instrument data are steadily increasing. A prime issue is that high volume data, produced by the next generation of earth observation instruments, cannot be transmitted to earth efficiently, since science data downlinks only offer limited capacity. Therefore, novel approaches to onboard processing are required. Utilizing reconfigurable hardware promises a significant improvement in the performance of onboard payload data processing. For this purpose, a system architecture has been developed that integrates avionic interfaces (e.g., SpaceWire and WizardLink) in combination with reconfigurable hardware for use by satellite payload systems [65]. SRAM-based FPGAs are used as the core components of this architecture and dynamic reconfiguration is utilized to exchange hardware modules for data processing at runtime in order to enable new or updated functionalities, for one. Secondly, dynamic reconfiguration is used to increase reliability by mitigating the above-mentioned radiation effects. In this context, dynamic reconfiguration can be subdivided into scheduled and event-driven reconfiguration, as in *D.M.f.I.T.S*, [55], Sect. 5.4.5.2. Scheduled reconfiguration can be used to implement resource-sharing on hardware level by utilizing time-division multiplexing across the available FPGA area. Furthermore, scheduled reconfiguration can be used to minimize radiation effects by employing continuous scrubbing, in which a continuous readback of the configuration data is combined with an online check of the integrity of the data as it is obtained. While scrubbing is performed periodically with a period (scrubbing rate) in the range of seconds or milliseconds, scheduled reconfiguration can also be used to execute maintenance functions which are executed in longer time intervals. In our implementation, the scrubbing rate is limited solely by the clock frequency of the FPGA's configuration interface. The Xilinx Virtex-4 FPGAs used in the proposed satellite payload processing system offer a maximum configuration rate of 400 MB/s, resulting in a maximum scrubbing frequency of 100 Hz when scrubbing the complete FPGA. Higher frequencies are possible if only certain sections of the FPGA are scrubbed. The priority of each scrubbing task is a direct result of the defined reconfiguration schedule.

In event-driven reconfiguration, the loading and unloading of hardware modules is triggered by events. This could mean mission events, where, for instance, a sensor has detected an object of interest which requires further investigation using new hardware modules. It could, however, also be failure events, in which parts of the system fail to operate correctly; in this case, reconfiguration is used for self-healing by restoring the hardware configuration in an error-free area of the device.

Faults should not only be detected (and corrected), but should also be monitored at runtime; therefore, all monitored data are analyzed and the results are used as additional control inputs for the system. Using self-optimization, the system can be reconfigured to a very safe mode (i.e., high redundancy but low performance) if the error rate increases, while it can operate with higher performance and less redundancy if low error rates are detected.

The proposed scheme of adaptive redundancy can also be implemented on a more fine-grained level. Dynamic reconfiguration, as described above, will enable the implementation of different levels of redundancy and reliability for the partially reconfigurable hardware modules. As mentioned above, modules representing components where faults are unacceptable are implemented using very safe but space-consuming techniques such as TMR. Additionally, since TMR is a modular approach, it can be efficiently combined with our approach for dynamic reconfiguration. In this case, redundancy is adapted at run-time by inserting additional module instances, as well as the required majority-voting system. Other PR modules may implement components where faults are acceptable; however, these faults have to be detected immediately (e.g., communication using automatic repeat request). Moreover, PR modules may exist where faults are both acceptable, and can be detected or corrected later.

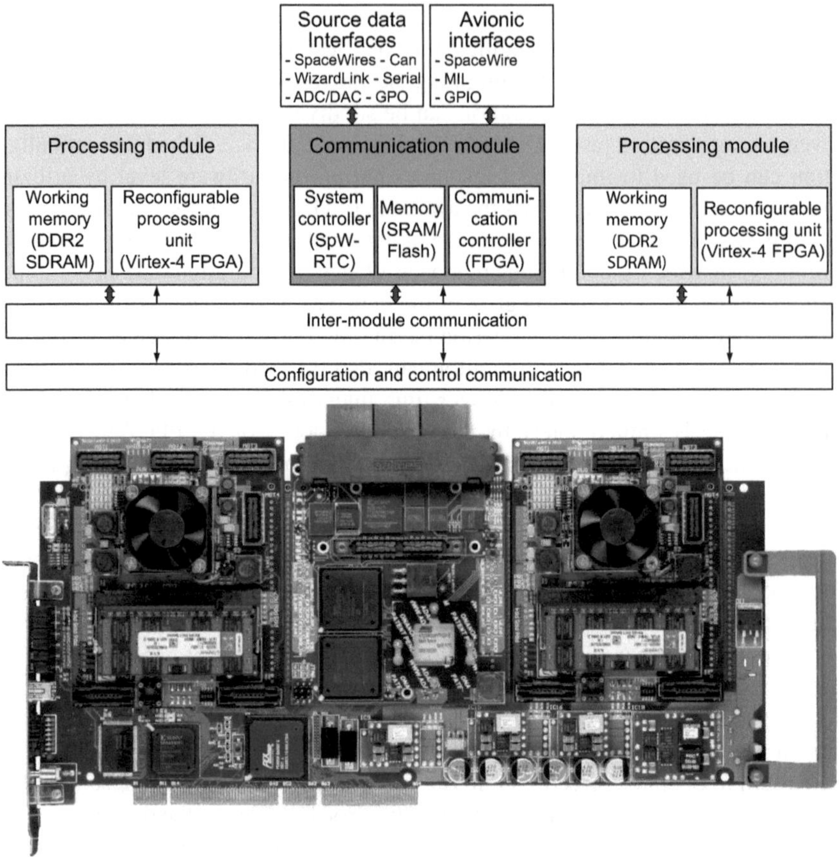

Fig. 3.52 Overview of a scalable system architecture for a payload processing system and the realization on the RAPTOR prototyping system

For validation of the proposed concepts, a hardware implementation of a self-optimizing satellite payload processing system with self-healing capabilities has been created based on the RAPTOR [117] prototyping system, as depicted in Fig. 3.52. The architecture can be easily scaled by the integration of additional Processing Modules integrating dynamically reconfigurable resources (Xilinx Virtex-4 FX100 FPGA) and by additional Communication Modules providing the required avionic and source data interfaces. The System Controller, a SpaceWire-RTC which consists of a LEON2-FT CPU, two SpaceWire interfaces and additional source data interfaces, controls the communication between the modules.

3.2.14 Online Model Checking

Franz-Josef Rammig and Yuhong Zhao

As we have already seen several times in preceding chapters, self-optimizing systems are capable of adjusting their behavior at runtime in response to changes in application objectives and the environment. Ensuring the dependability of such dynamic systems has given rise to new demands on verification techniques. The dynamic nature of these systems makes it difficult for traditional model checking techniques [30] to investigate the state space of the system model offline with reasonable time and space consumption. To overcome this problem, many efficient techniques have been presented in the literature so far: partial order reduction [91], compositional reasoning [18], abstraction technique [29] and bounded model checking [20], to name just a few. These improvements enable model checking to verify more complex systems, unfortunately, however, at the cost of making the model checking process more complicated. Self-optimizing systems further exacerbate the state-space explosion problem. Using the above-mentioned verification techniques to check self-optimizing systems offline still leaves much to be desired.

Traditional testing [104] provides a partial proof of correctness at the level of system implementation. For untested inputs, undiscovered errors in deep corners might show up during system execution.

Runtime Verification [10, 14, 28, 40, 41, 68, 143] also works on a system implementation level. It attempts to check the correctness of the sequence of states monitored or derived from the current execution trace. Runtime verification can proceed further only after a new state has been observed. Therefore, it is difficult to detect errors before they have already occurred.

Online Model Checking [154] works on the system implementation level as well, but it checks the correctness of the corresponding system model. Errors at the model level might indicate potential errors at the implementation level. Simply speaking, online model checking is a lightweight verification technique to ensure at runtime the correctness of the current execution trace of the system application under test by means of checking a partial state space of the system model covering the execution trace. For this purpose, we need to monitor the system execution trace from time to time; in doing so, we can avoid the state-space explosion problem. The observed (current) states are used to locate the partial state space to be explored. Online model checking aims to "look" into the near future in the state space of the system model, in order to see whether potential errors are lurking there or not. As a side effect, the conformance of the implementation to the corresponding model can also be checked during the process. The counterexample provided by online model checking is a clue to help locate the error(s), which might be in a deep corner and thus hard to reproduce.

Notice that we do not directly check the actual execution trace itself; thus, the progress of our online model checking is not tightly bound to that of the system execution. This means that, if we can make online model checking sufficiently efficient, it is possible to predict potential errors before they have actually occurred. The efficiency of online model checking depends primarily on the search algorithm and the underlying hardware architecture, as well as on the complexity of the checking problem. In the following, we present our efficient online model checking mechanism and its implementation as a system service of a Real-time Operating System.

3.2.14.1 Prerequisites and Input

Our online model checking mechanism is integrated into a Real-time Operating System called ORCOS (Organic ReConfigurable Operating System [48]) as a system service. The system application (source code) under test and its formal model, as well as the property to be checked, are known in advance to ORCOS. The formal model is derived from the system specification or extracted from the system implementation. The property is given in form of an invariant or of a general Linear Temporal Logic (LTL) formula.

3.2.14.2 Description

Online model checking attempts to check (at the model level) whether the current execution trace (at the implementation level) could run into a predefined unsafe region (error states) or not.

Basic Idea

Without loss of generality, suppose that the current state s_i of the system application under test can be monitored in some way from time to time and that the unsafe region is derived offline from the property to be checked. Fig. 3.53 illustrates the basic idea of our online model checking mechanism.

For each checking cycle, whenever a new current state s_i is monitored during system execution, we can use the corresponding abstract state $\widehat{s_i} = \alpha(s_i)$, where the function $\alpha(\cdot)$ maps a concrete state s_i to an abstract state $\widehat{s_i}$, to reduce the state

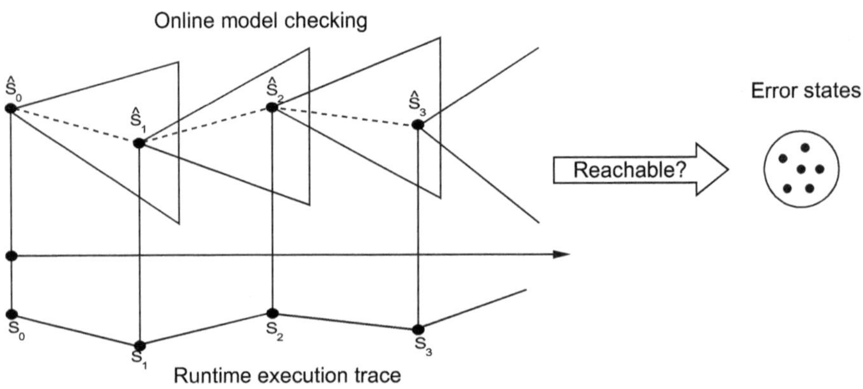

Fig. 3.53 Online model checking

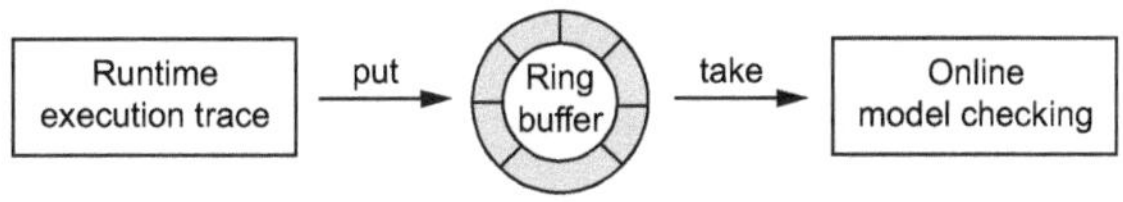

Fig. 3.54 Communication mechanism

space to be explored by online model checking. It is therefore is sufficient to explore a partial state space starting from the corresponding abstract state $\widehat{s_i}$. It is worth mentioning that, if no abstract state is consistent with $\alpha(s_i)$, it means that the implementation of the system application does not conform to its model. This consistency checking is a byproduct of online model checking. Because of the limited checking time allocated to online model checking, for each checking cycle, only a finite number of transition steps, say the next k steps, starting from the observed state will be explored. Therefore, online model checking is, in essence, Bounded Model Checking (BMC) [20] applied at runtime.

An SAT solver can be used as a verification engine for online model checking. In case of a relatively small k, [20] concludes that "SAT based BMC is typically faster in finding bugs compared to BDDs". Unfortunately, [20] also concludes that "The deeper the bug is, the less advantage BMC has." However, by doing BMC at runtime, it is quite possible to find deep corner bugs (if any) in the state space of a large complex system. If no error is detected, then the execution trace is safe for at least the next k steps. Once an error has been detected, online model checking will inform the underlying operating system in time. It is then up to the operating system to decide how to deal with the particular case.

Communication Mechanism

In order to obtain current state information, we can have our online model checking communicate with the system application through a ring buffer, as shown in Fig. 3.54. A special monitor, which can observe the current state while the system application is running, puts the states into the buffer from time to time, while the online model checking periodically tries to take a state from the buffer. This state is used to decide a partial state space of the system model to be explored by the online model checking. If the buffer is full, the oldest state is overwritten by the latest one. It is easy to see that the progress of the online model checking is not strongly bound to the system execution due to its working at the model level. Online model checking does not check the actual execution trace itself; it merely uses the monitored states to reduce the state space to be explored. This is different from state-of-the-art runtime verification, which checks the observed execution trace itself.

Accelerating Online Model Checking

In practice, online model checking could run ahead of or fall behind the execution of the system application. In the former case, it is possible for online model checking to predict the potential errors before they've actually occurred. Therefore, it would be advantageous to speed up online model checking [125], so that we have a better chance to fulfilling this goal. One possibility would bet to speed up the search

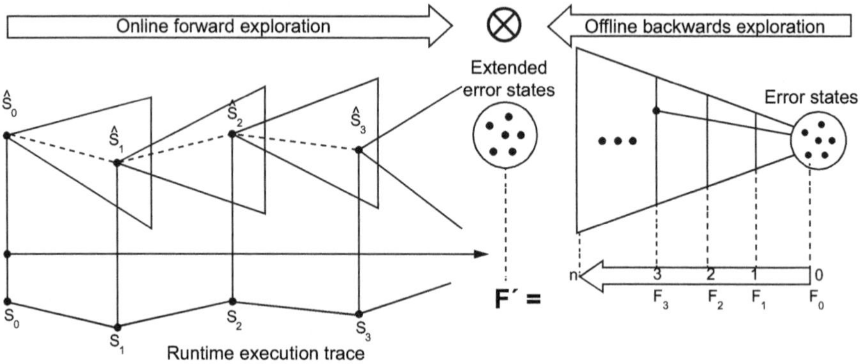

Fig. 3.55 Accelerating online model checking

algorithm by using an efficient SAT solver optimized and customized for online model checking. The second possibility is to reduce the workload of online model checking by introducing offline backward exploration, as shown in Fig. 3.55.

For this purpose, we need to derive (initial) unsafe condition from the LTL formula to be checked. In case of a safety property, it is trivially a reachability problem, i.e., the error path is finite; the unsafe condition can be obtained easily by means of a negation operation. However, in case of a liveness property, the error path is infinite. For finite state systems, this means, the error path must end at some accepting state that satisfies the fairness condition and that has some loop back to it in the meantime. Online checking of a loop condition will substantially increase the workload of online model checking. Fortunately, we can always obtain the fairness condition in advance from the Büchi automaton derived from the liveness property. Therefore, we can calculate offline a set of states that satisfy the fairness condition and that have some loop back to them. This set of states can be seen as the (initial) error states.

Now we can extend the (initial) unsafe region F_0 to become $F' = F_0 \vee F_1 \vee \cdots \vee F_n$ by offline backward exploration up to n time steps, as shown in Fig. 3.55. As a result, online model checking is reduced to online reachability checking [125], a simple form of Bounded Model Checking.

Many existing efficient solutions to traditional model checking can be directly applied to offline backward exploration. Given enough time and memory, it is possible to explore backwards much deeper in the state space of the system model to be checked. Doing so will thus substantially reduce the workload of online forward exploration.

3.2.14.3 Results

The existence of a Real-time Operating System (RTOS) implies that any effect of a closed-loop control system, be it sending a control value to or receiving an input

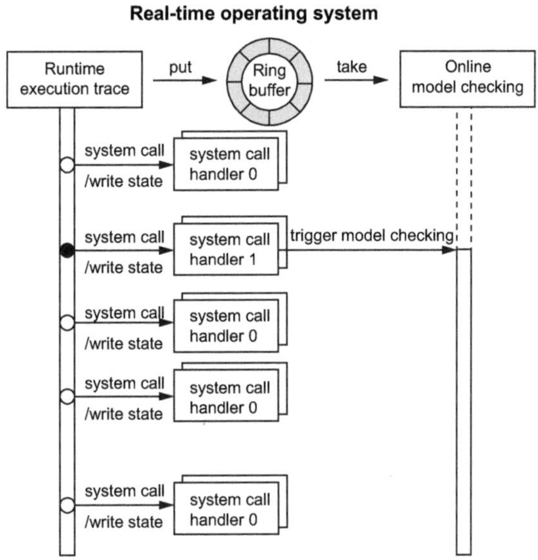

Fig. 3.56 Underlying system architecture

value from the controlled object or be it any kind of communication with another application task, happens under the control (and this includes notion) of the RTOS. All such actions of an application can happen only by means of system calls. A failure can be malign according to Kopetz [84] only when passed to the outside via a system call, while all other failures are benign.

That is, the implementation of any such application has to contain a sequence of system calls. Whenever a system call is invoked, we can monitor the state information of the implementation. Therefore the sequence of system calls of an application is the appropriate level of granularity, at which we can monitor state information used by online model checking. Such system calls happen anyhow during the system execution. Online verification can thus be integrated as part of the system call handler of an RTOS, thereby causing no additional context switch overhead and with the necessary information already available without crossing address space borders. We assume that there exist two versions of a system call handler, one with integrated online model checking and one without. When entering a critical application or critical part of an application, then it just means to switch to the proper mode of system calls. In this sense online model checking becomes an RTOS service, as shown in Fig. 3.56.

3.2.14.4 Application Example

Nowadays industries rely increasingly on (embedded) *software* for their product safety. For instance, the vehicles' electronics systems are usually controlled by software with millions of lines. Online model checking mechanism can increase the dependability of the safety-critical systems to some degree. If an error is detected in

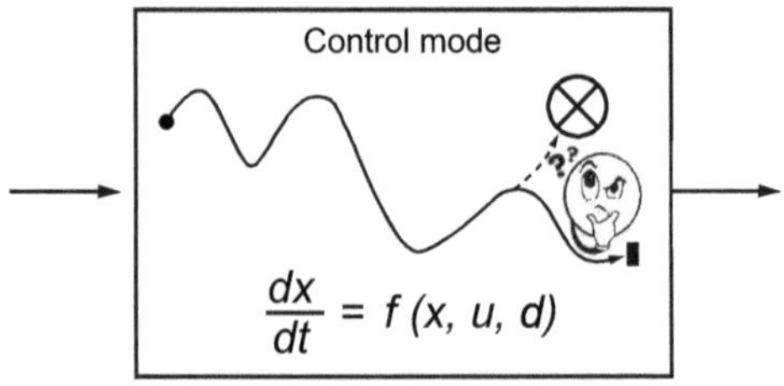

Fig. 3.57 Online model checking problem for hybrid systems

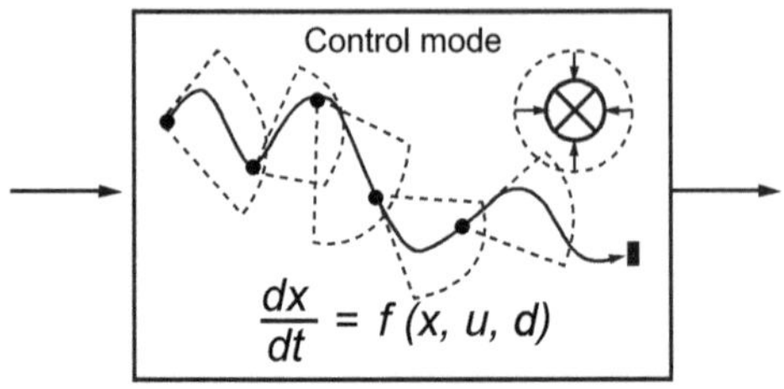

Fig. 3.58 Online hybrid reachability checking

time, the underlying operating system then has time to react to the error. At the very least, the counterexample provided by online model checking can help the user to figure out the location of and reason for the subtle error, which is usually difficult to reproduce in a laboratory environment for the large complex software systems.

On the other hand, our online model checking mechanism can also be used to increase the safety of the *hybrid* systems as shown in Fig. 3.57, where the symbol $\otimes$ represents the unsafe region of the continuous state space associated with the given control mode. Starting from the unsafe region, we can calculate offline a backward reachable set up to n time steps. At runtime, the system states are sampled from time to time. Then, we can check online, as shown in Fig. 3.58, whether the trajectory from each observed state could reach the extended unsafe region in the near future or not.

3.2.15 *Virtualization*

Stefan Groesbrink

Self-optimizing mechatronic systems are in many cases characterized by higher demands on processing power and memory compared to traditional control systems. Due to space constraints, merely increasing the number of processing units and the resulting extension of the cable infrastructure is often not feasible. Virtualization can reconcile these opposing requirements and increase the processing power without increasing the space. It achieves this by providing multiple execution environments, which enable the consolidation of multiple systems onto a single hardware platform (system virtualization). Instead of adding control units, more powerful control units increase the system's processing power. Virtualization's architectural abstraction supports the migration from single-core to multi-core platforms and helps to utilize this multi-core hardware efficiently. Multi-core platforms enable higher processing performance at lower electrical power per frequency, resulting in less heat dissipation.

Mechatronic systems are typically safety-critical. Therefore, it is of paramount importance that the integrated subsystems do not interfere with each other. Virtualization's integration of multiple systems does not lead to a loss of isolation; both spatial and temporal separation can be maintained (brick wall partitioning). By consequence, independently developed software components, such as third party components, trusted legacy software, and newly developed application-specific software, can be combined to achieve the required functionality. The reusability of software is increased without endangering reliability and safety.

In addition to fostering safety-related characteristics, the application of virtualization can actively improve two major attributes of dependability, namely reliability and availability. The virtual machine concept, with its encapsulation of a subsystem's state, supports migration; by enabling the migration of virtual machines, a system can respond to unforeseen failures at runtime. In case of a partially failed processor unit, partial memory failures, or a breakdown of acceleration coprocessors, the operation of the subsystem can be continued on another processor, as long as it is still possible to save and transfer the virtual machine's state. One particular benefit is self-diagnosing hardware that signals upcoming hardware failures on the basis of built-in self-tests.

Virtualization is an architectural measure to improve safety, reliability and availability. In addition to this, by applying the methods *Analysis of Self-Healing Operations* (Sect. 3.2.8) and *Online Model Checking Mechanism* (Sect. 3.2.14) , the system behavior can be monitored at runtime in order to identify malicious system states.

3.2.15.1 Prerequisites and Input

The application of system virtualization requires a software layer that virtualizes the hardware resources in order to provide multiple execution environments; this system software component is called a hypervisor or virtual machine monitor. The hypervisor has to be configured for each specific application according to the characteristics of the subsystems consisting of operating system and applications, which are to be consolidated. For paravirtualization (modification of the operating system to be aware of the fact that it is executed on top of a hypervisor) neither technical nor legal issues should preclude the modification of the operating system source code.

3.2.15.2 Description

System virtualization describes the methodology of dividing the resources of a complete computer system into multiple execution environments (platform replication) [139]. The underlying physical hardware can be shared among multiple operating system instances, even among different operating systems, in order to

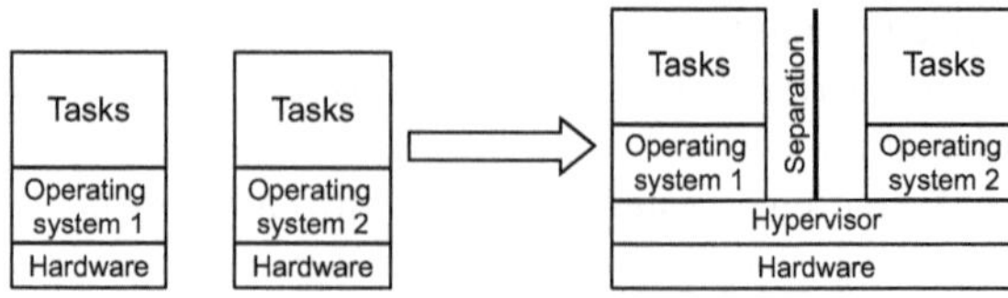

Fig. 3.59 System virtualization

provide each software component with a suitable system-software interface (see Fig. 3.59).

The hypervisor (also known as *virtual machine monitor*) implements the virtualization layer. It creates and manages *virtual machines*, also referred to as partitions, which are isolated conceptual duplicates of the real machine [115]. Guest systems are executed within these virtual machines. The real machine is the hardware environment, including processor, memory, and I/O resources, with the instruction set architecture as the interface for the system software. A virtual machine does not have to be characterized by exactly the same hardware environment as the real machine. For example, the instruction set architectures might not be identical, in which case the hypervisor maps the instruction set of the virtual machine onto the instruction set of the physical machine. Some physical resources, such as memory, can be partitioned so that each virtual machine uses a certain fraction. This is not possible for other resources, for example the processor in a uniprocessor system, in which case time-sharing has to be applied [139].

The hypervisor retains control of the real hardware resources, without exception. If a resource is made available to multiple virtual machines, the illusion for the operating systems of having exclusive access to this resource is maintained by the hypervisor. When an operating system performs a *privileged instruction*, i.e. an instruction that directly accesses the machine state, the hypervisor intercepts the operation. If an operating system tries, for example, to set a control flag of the microprocessor, this cannot be allowed, since the modification would influence the behavior of the other operating systems and destroy their illusion. Therefore, the hypervisor intercepts this privileged instruction, stores the current value of the flag and sets the control flag, but resets it before it performs a virtual machine switch.

Hypervisors are classified by their capability to host unmodified operating systems. In terms of *full virtualization*, unmodified operating systems can be executed within a virtual machine. In contrast, paravirtualization requires a porting to the hypervisor's paravirtualization application programming interface [13]. The operating system is aware of being executed within a virtual machine and can use hypercalls to request hypervisor services. Paravirtualization can often be exploited to increase the performance [78]; however, the major drawback is the need to perform modifications of critical kernel parts of an operating system. If legal or technical issues preclude this, for example in case of a commercial operating system, it is not possible to host it.

The application of system virtualization to embedded real-time systems requires guaranteeing *spatial and temporal separation* of the hosted guest systems. Spatial separation refers to protecting the integrity of the memory spaces of both the hypervisor and the guests. Any possibility of a harmful activity going beyond the boundaries of a virtual machine has to be precluded. Spatial separation can be ensured by hardware components such as memory management units or memory protection units, which are available for many embedded processors. Temporal separation is fulfilled, if all guest systems are executed in compliance with their timing requirements, meaning that a predictable, deterministic behavior of every single real-time guest must be guaranteed. System virtualization implies scheduling decisions on two levels (hierarchical scheduling): the hypervisor schedules the virtual machines and the hosted operating systems schedule their tasks with their own schedulers. The hypervisor has the responsibility of scheduling the virtual machines in a manner that assigns the subsystems early enough and gives them enough time to complete their computations on schedule.

3.2.15.3 Results

Proteus [12, 61] is a real-time hypervisor for multi-core PowerPC platforms. The software design (depicted in Fig. 3.60) is based on the *Multiple Independent Levels of Security (MILS)* approach for highly robust systems [7]. According to this design concept, a system is divided into multiple isolated components, consisting of program code, data, and system resources, with no way for information to flow except through the defined paths. Only the minimal set of components runs in supervisor mode (privileged mode): interrupt and hypercall handlers, the virtual machine scheduler, and the inter-partition communication manager (IPCM), which is responsible for communication between virtual machines. All other components, such as

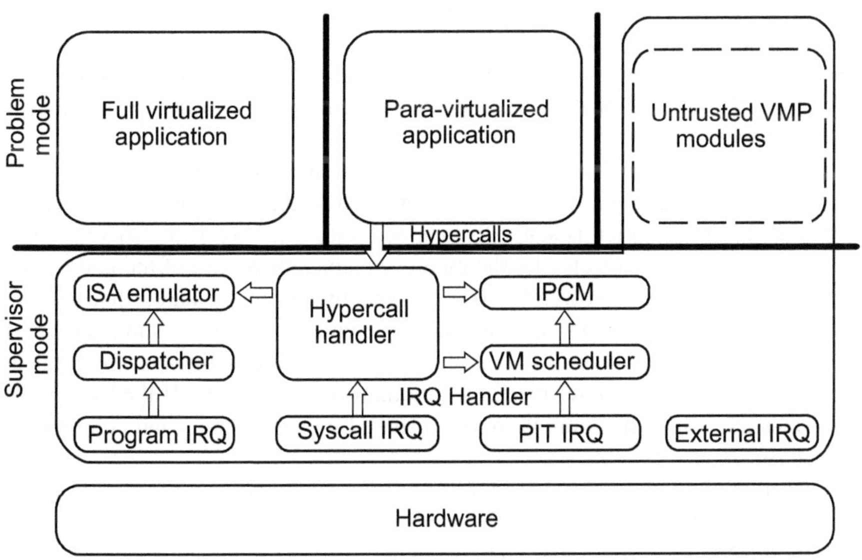

Fig. 3.60 Design of the Proteus hypervisor

I/O device drivers, are placed inside a separate partition (Untrusted VMP Modules) and executed in problem mode (user mode). The interrupt handling is the central component of the Proteus architecture. Any occurring interrupt is delegated to the hypervisor, which saves the context of the running virtual machine and forwards the interrupt request to either the responsible component or back to the guest operating system.

Proteus is a symmetrical hypervisor with no distinction between cores; all cores execute guest systems. When required, for example in case of a guest's call for a hypervisor service, the hypervisor takes control and its own code is executed on that core. Different guests on different cores can perform this context switch, from guest to hypervisor, simultaneously.

As a *bare-metal* hypervisor, Proteus runs directly on top of the hardware, without an underlying host operating system [139]. Controlling the hardware directly facilitates a more efficient virtualization solution. Resource management and especially scheduling is not at the mercy of a host operating system; the amount of code executed in privileged mode is smaller; and no operating system is incorporated in the trusted computing base, which increases the overall security and the certifiability of functional safety.

Proteus features both paravirtualization and full virtualization. Paravirtualization can be exploited to increase the efficiency, but is characterized by a limited applicability. The support of non-modifiable guests requires full virtualization. In addition, tasks without underlying operating system can be executed on top of the hypervisor. The concurrent hosting of any combination of paravirtualized guests, fully virtualized guests, and guest tasks without an operating system is possible unrestrictedly.

The paravirtualization interface is characterized by two main functionalities. In addition to providing a handler routine for each privileged instruction which emulates the instruction, the paravirtualization interface offers additional services. Paravirtualized operating systems can communicate with other guests through *hypercalls*, call I/O functionality, pass scheduling information to the hypervisor, or yield the CPU.

As mentioned above, Proteus guarantees spatial and temporal separation of the guest systems. To achieve spatial separation, each virtual machine operates in its own address space, which is statically mapped onto a region of the shared memory. This mapping is protected by the memory management unit of the underlying hardware platform (e.g. PowerPC 405). If systems are consolidated that have to communicate with each other, it is mandatory to ensure that this communication is still possible. The only path of data flow between virtual machines is communication via the hypervisor's inter-partition communication manager. If the hypervisor authorizes the communication, it creates a shared memory tunnel and controls the correct use.

Ensuring temporal separation requires multiple methods. As a precondition, the worst-case execution times of all hypervisor routines were determined by path analysis of the executable files. The knowledge of these bounded execution times make it possible to determine the worst-case execution time of a program executed on top of Proteus. Each virtual machine can be executed on any core; if this is undesired, a virtual machine can be assigned to one specific core or a subset of cores; for example, a core exclusively to a safety-critical guest. If multiple virtual machines are assigned to one core, they share it in a time-division multiplexing manner. The virtual machine scheduling is implemented based on fixed time slices. The guests' task sets have to be analyzed and execution time slots within a repetitive major cycle are assigned to the virtual machines, based on the required utilization and execution frequency. A real-time response time conserving calculation of the time slots was developed [76].

Access to shared resources, such as peripheral devices, has to be synchronized. One common solution is the use of semaphores assigned exclusively to one core at any time. The PowerPC 405 does not feature any hardware support for mutual exclusion in a multi-core architecture, which is why Proteus uses a software implementation that does not explicitly rely on hardware support: Leslie Lamport's Bakery Algorithm [88]. It was selected since it does not require atomic operations, such as compare-and-swap or test-and-set; it also satisfies FIFO fairness and excludes starvation. The software synchronization mechanism extends the applicability of the hypervisor to shared memory multiprocessor platforms. The processors work on their own random access memory, but are connected by a bus hierarchy and can use shared memory. In such a multiprocessor system, virtual machines can be relocated through migration from one processor to another one. We developed analytical means to evaluate whether a migration of a virtual machine with real-time constraints can be performed without risking a deadline miss [63].

By the application of processor emulation, Proteus supports even heterogeneous multiprocessor platforms, which are characterized by processors with differing instruction set architectures. Heterogeneous platforms provide suitable processors for different applications, such as general processing, network control, or signal processing. The combination of emulation and heterogeneous platforms enables the consolidation of legacy systems that were developed for different architectures. Emulation enables the execution of program binaries that were initially compiled for a different architecture; it translates between instruction set architectures and therefore makes cross-platform software portability possible. With the use of emulation techniques, Proteus supports the migration of virtual machines even from one processor to one processor with a different instruction set architecture at runtime [62]. We developed an emulation approach that maintains real-time capability [77]; it minimizes the worst-case execution time overhead and combines interpretation and binary translation for an optimal trade-off between required memory and performance.

In a pre-processing step, critical basic blocks are identified; these blocks are characterized by a high performance ratio between the emulated execution on the host as compared to native execution. These blocks are translated once and the result is stored, meaning that the size of the available memory limits this optimization. Interpretation is used for non-selected blocks.

Proteus offers static configurability. Depending on the requirements of the actual system, the system designer can configure the hypervisor by modifying a configuration file. According to these specifications, the preprocessor manipulates the implementation files and removes unneeded code. The system designer can decide to enable TLB virtualization, device driver support, inter-partition communication and multiple performance-enhancement features.

3.2.15.4 Application Example

The basic idea of system virtualization is to create an integrated system that combines the functionality of multiple systems in order to attain a complex behavior. This modular synthesis has the potential to reduce development time by increasingly reusing trusted systems. The consolidation on a single hardware platform can often provide more efficient implementations as concerns regarding power consumption, hardware footprint and system costs. Multiple operating systems can be hosted to provide all applications with a suitable interface. Industrial automation systems, for example, often require both a deterministic real-time operating system for the control of actuators and a feature-rich general-purpose operating system for the human-machine interface and connection to the corporate IT. Of paramount importance from a dependability point of view, system virtualization maintains the isolation between the integrated systems. Spatial separation precludes any possibility of a harmful activity going beyond the boundaries of a virtual machine and temporal separation precludes a guest not being able to meet its timing requirements due to interference from another guest.

3.3 Methodology for the Selection of Dependability Methods for the Development of Self-optimizing Systems

Rafal Dorociak, Jürgen Gausemeier, and Peter Iwanek

The development of dependable self-optimizing systems is a difficult task that engineers have to face, involving many complex aspects such as non-deterministic behavior and cross-domain fault propagation. There is also an immense number of engineering methods which can be used to improve the dependability of such self-optimizing systems. Many of these specifically intend to improve the dependability of self-optimizing systems have already been discussed in this chapter, such as Iterative Learning Stochastic Disturbance Profiles in Sect. 3.2.2, or Online Model Checking in Sect. 3.2.14. Beyond the scope of this book, there are also many more methods which can be used during the development phase of technical systems, such as the FTA, FMEA etc. (some of the existing databases provide information on more than 700 such methods [105]). Which of them are suitable for a particular system depends on the underlying development task and the Principle Solution of the system. The developer faces the challenge of how to choose suitable engineering methods from the vast number of available methods and how to embed them into the product development process. Nowadays, as a rule, the search for and selection of a method have to be done manually, which is tedious and often error-prone. Therefore, there is an evident need for a methodology which can aid developers in choosing and applying appropriate dependability engineering methods. Such a methodology has been developed within the Collaborative Research Center (CRC) 614 and is presented in the following. This methodology includes a method database, a guide to selection and planning the use of dependability engineering methods throughout the entire product development process, and a software tool. It enables the developer to choose and plan the use of suitable dependability engineering methods for the particular development task. The developer receives suggestions on which methods can be used, how they depend on each other, and how these methods can be combined, as well as what their optimal chronological order is. The end result of this process is a proposed workflow of process steps and methods. In the following, the constituent parts of the methodology are explained in more detail.

The *method database* contains the description of dependability engineering methods, which have been manually entered into the database. They are characterized by their inputs (e.g. specification of the Principle Solution) and outputs (e.g. description of the failure propagation) as well as by a number of criteria, such as the dependability attributs (e.g. safety), the domain (e.g. control engineering), the development phase (e.g. Conceptual Design), the industrial sector (e.g. automotive), relevant standards (e.g. CENELEC 50128 [46]), etc. Links to the development process and external documentation are also included. In addition, useful documents (e.g. templates, usage in former projects) and the relationships between methods are described as well. The following relationships between methods are those most frequently used in the methodology:

- "is a prerequisite for",
- "requires",
- "is the further development/continuation of",
- "has been further developed to" and
- "can be complemented by".

As a example: the *Early Probabilistic Reliability Analysis of an Advanced Mechatronic System based on its Principle Solution* (Sect. 3.1.1) "is a further development/continuation of" the *classic FTA* and *FMEA*. To use the *Early Probabilistic Reliability Analysis*, the specification of the system should be carried out using the specification technique CONSENS meaning that the *specification technique CONSENS* "is a prerequisite for" the *Early Probabilistic Reliability Analysis*. From another perspective, the *Early Probabilistic Reliability Analysis* "requires" the usage of the *specification technique CONSENS*. By showing these relationships between methods, the links between methods as well as their interactions can be clearly represented, providing an easily intelligible overview of how the methods are connected to one another.

Additionally, the methods can be classified with regard to their self-optimizing relevance as dependability engineering methods which are either specific to or which are not specific to self-optimizing systems:

Methods which are not specific to self-optimizing systems:
A number of classical dependability engineering methods can be used to improve the dependability of a self-optimizing system. Examples are FMEA, FTA, FHA, etc. They are usually performed on detailed system designs, and thus relatively late in the product development process. Some methods have been adapted for use in the Conceptual Design, such as the early FMEA method [38]. These classical methods allow initial statements with respect to the dependability of the system to be made. Based on those statements, potential weaknesses can be found in the Principle Solutions and counter-measures derived; one example of a possible counter-measure is the use of redundancy for a particular dependability-critical system element.

Methods which are specific to self-optimizing systems:
In addition, some dependability engineering methods have been developed within the CRC 614 which are specifically designed for use with self-optimizing systems (see also previous sections). They are usually used during operation i.e. the self-optimizing system is able to compensate for failures during operation and to alter its behavior to reach a dependable state. One example is the Multi-Level dependability concept (see Sect. 3.2.1), which integrates advanced condition monitoring [90, 140] into a self-optimizing system. Another example is a method for the Conceptual Design of the System of Objectives for a self-optimizing system [114].

The above-mentioned classification systems have been incorporated into our methodology for aiding developers during the improvement of the dependability of self-optimizing systems.

A *software tool* has also been developed for use in this methodology. It supports the insertion of new and the modification of existing method descriptions and offers a search function to make it easier for the user to find suitable methods. It allows users to search for dependability engineering methods with regard to their characterization criteria (see Chap. 4 for a more detailed example).

The *guide for planning methods* proposes in which sequence the selected dependability engineering methods should be used. Methods will be selected by the user based on the list of methods returned by the software tool. This recommendation is based on two pieces of information, both of which are stored in the database: The first piece of information is the description of the input and output relationships between the methods, and the development tasks and their sequences, which are defined for each of the methods. These sequences are also saved in the database's method description; if the user selects one of the displayed methods, the corresponding sequence will be presented.

In the following, a brief description of how the methodology can be used is provided: First, a search for specific dependability engineering methods is performed using the method database (Step 1 in Fig. 3.61). The result is a list of all recommended methods which match the search criteria. As stated before, these search criteria (corresponding domain, relevant dependability attributes, respective development phase, relevant standards, etc.) are provided by the safety engineer

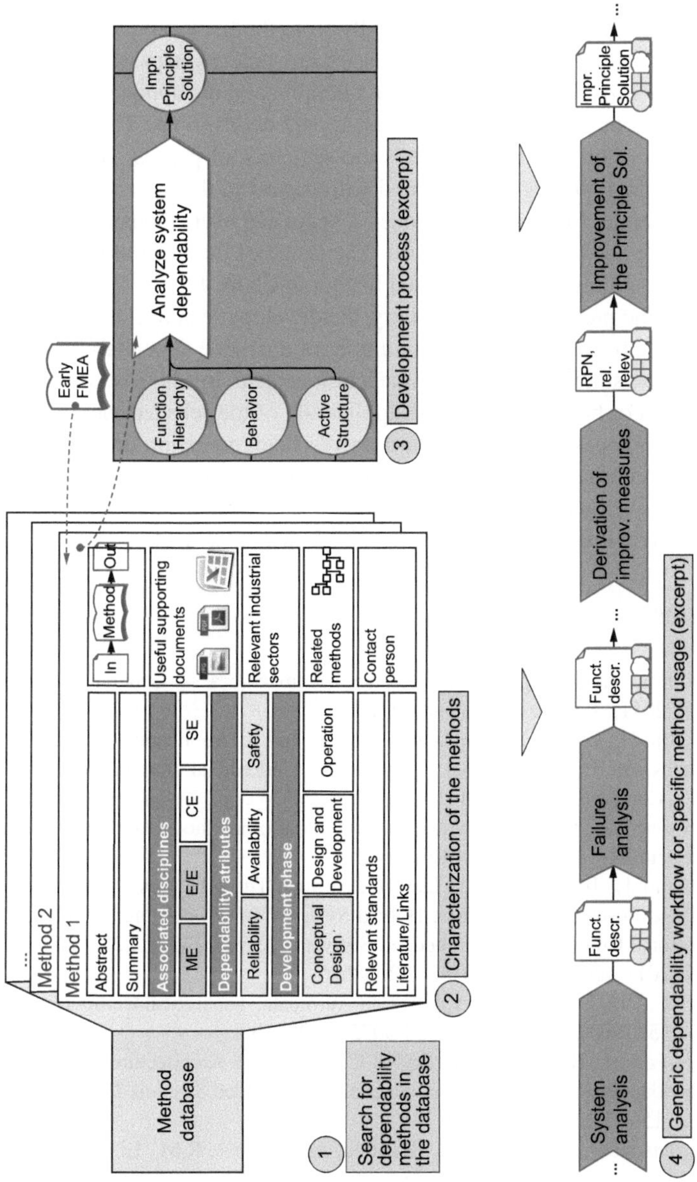

Fig. 3.61 Selection and planning of dependability engineering methods with regard to the underlying development task

and correspond to the development task at hand. The user selects the appropriate dependability engineering methods manually from the list (Step 2). From the method database, the user navigates to the corresponding process steps in the process model (Step 3). The software tool supports the planning and the application of the chosen methods based on the underlying process model descriptions. The result is a workflow diagram (Step 4) which suggests the sequence of process steps and methods to be used. The planning is performed with regard to the underlying development task; for example, for a safety engineer, a sequence of methods is proposed which complies with a given safety standard. The usage of the methodology and the corrosponding methods will be discussed further in Chap. 4.

By using the presented methodology, the developer can decide both more easily and more quickly which of the vast number of available dependability engineering methods is best suited to the development task at hand. In addition, this methodology can assist him or her in planning the application of the selected methods during the development process. It also encourages the documentation of best practice, i.e. exemplary combinations of methods used in projects which can be repeated later. New projects can later be structured upon those best practices.

The concrete use of this methodology is shown in Chap. 4.

References

1. Condition monitoring and diagnostics of machines – General guidelines (ISO 17359:2011). International Standard (2011)
2. Aickelin, U:, Cayzer, S.: The Danger Theory and Its Application to Artificial Immune Systems. In: 1st International Conference on ARtificial Immune Systems (ICARIS 2002), Canterbury, UK, pp. 141–148 (2002)
3. Allgower, E.L., Georg, K.: Numerical continuation methods, vol. 33. Springer, Berlin (1990), doi:10.1007/978-3-642-61257-2
4. Alpaydın, E.: Introduction to Machine Learning. The MIT Press (2004)
5. Alur, R.: Formal Verification of Hybrid Systems. In: Proceedings of the 9th ACM International Conference on Embedded Software, Taipei, TW, pp. 273–278. ACM, New York (2011), doi:10.1145/2038642.2038685
6. Alur, R., Dill, D.L.: A Theory of Timed Automata. Theoretical Computer Science 126, 183–235 (1994), doi:10.1016/0304-3975(94)90010-8
7. Alves-Foss, J., Harrison, W.S., Taylor, C.: The MILS Architecture for High Assurance Embedded Systems. International Journal of Embedded Systems 2(3), 239–247 (2006), doi:10.1504/IJES.2006.014859
8. Androutsopoulos, K., Clark, D., Harman, M., Hierons, R.M., Li, Z., Tratt, L.: Amorphous Slicing of Extended Finite State Machines. IEEE Transactions on Software Engineering 99(PrePrints), 1 (2012), doi:10.1109/TSE.2012.72
9. Anis, A., Goschin, S., Lehrig, S., Stritzke, C., Zolynski, T.: Developer Documentation of the Project Group SafeBots II. Project group. University of Paderborn, Department of Computer Science, Paderborn, DE (2012)
10. Arkoudas, K., Rinard, M.: Deductive Runtime Certification. In: Proceedings of the 2004 Workshop on Runtime Verification (RV 2004), Barcelona, ES (2004), doi:10.1016/j.entcs.2004.01.035

11. Avizienis, A., Laprie, J.C., Randell, B., Landwehr, C.: Basic Concepts and Taxonomy of Dependable and Secure Computing. IEEE Transactions on Dependable and Secure Computing 1(1), 11–33 (2004), doi:10.1109/TDSC.2004.2
12. Baldin, D., Kerstan, T.: Proteus, a Hybrid Virtualization Platform for Embedded Systems. In: Rettberg, A., Zanella, M.C., Amann, M., Keckeisen, M., Rammig, F.J. (eds.) IESS 2009. IFIP AICT, vol. 310, pp. 185–194. Springer, Heidelberg (2009)
13. Barham, P., Dragovic, B., Fraser, K., Hand, S., Harris, T., Ho, A., Neugebauer, R., Pratt, I., Warfield, A.: Xen and the Art of Virtualization. In: Proceedings of the 19th ACM Symposium on Operating Systems Principles, Bolton Landing, NY, US (2003), doi:10.1145/945445.945462
14. Barnett, M., Schulte, W.: Spying on Components: A Runtime Verification Technique. In: Leavens, G.T., Sitaraman, M., Giannakopoulou, D. (eds.) Workshop on Specification and Verification of Component-Based Systems, pp. 1–9. Published as Iowa State Technical Report 01-09a (2001)
15. Becker, S., Brenner, C., Brink, C., Dziwok, S., Heinzemann, C., Löffler, R., Pohlmann, U., Schäfer, W., Suck, J., Sudmann, O.: The MechatronicUML Design Method – Process, Syntax, and Semantics. Tech. Rep. tr-ri-12-326, Software Engineering Group. Heinz Nixdorf Institute, University of Paderborn (2012)
16. Behrmann, G., David, A., Larsen, K.G., Pettersson, P., Yi, W., Hendriks, M.: Uppaal 4.0. In: Quantitative Evaluation of Systems, QEST 2006, pp. 125–126. IEEE Computer Society (2006), doi:10.1109/QEST.2006.59
17. Ben-Gal, I.: Bayesian Networks. Encyclopedia of Statistics in Quality and Reliability (2007), doi:10.1002/9780470061572.eqr089
18. Berezin, S., Campos, S.V.A., Clarke, E.M.: Compositional Reasoning in Model Checking. In: de Roever, W.-P., Langmaack, H., Pnueli, A. (eds.) COMPOS 1997. LNCS, vol. 1536, pp. 81–102. Springer, Heidelberg (1998)
19. Bielawny, D., Krüger, M., Reinold, P., Timmermann, J., Trächtler, A.: Iterative learning of Stochastic Disturbance Profiles Using Bayesian Networks. In: 9th International Conference on Industrial Informatics (INDIN), Lisbon, PT (2011), doi:10.1109/INDIN.2011.6034920
20. Biere, A., Cimatti, A., Clarke, E.M., Strichman, O., Zhu, Y.: Bounded model checking. Advances in Computers 58, 118–149 (2003), `http://repository.cmu.edu/compsci/451`
21. Birolini, A.: Reliability Engineering – Theory and Practice, 5th edn. Springer, Heidelberg (2007), doi:10.1007/978-3-662-03792-8
22. Blanke, M., Kinnaert, M., Lunze, J., Staroswiecki, M.: Diagnosis and Fault-Tolerant Control. Springer (2006), doi:10.1007/978-3-662-05344-7
23. Blesken, M., Rückert, U., Steenken, D., Witting, K., Dellnitz, M.: Multiobjective optimization for transistor sizing of CMOS logic standard cells using set-oriented numerical techniques. In: NORCHIP 2009, pp. 1–4 (2009), doi:10.1109/NORCHP.2009.5397800
24. Borkar, S.: Designing Reliable Systems from Unreliable Components: The Challenges of Transistor Variability and Degradation. IEEE Micro 25(6), 10–16 (2005), doi:10.1109/MM.2005.110
25. Cao, Y., Hussaini, M., Zang, T.: An Efficient Monte Carlo Method for Optimal Control Problems with Uncertainty. Computational Optimization and Applications 26, 219–230 (2003), doi:10.1023/A:1026079021836
26. Cassez, F., Roux, O.H.: Structural Translation from Time Petri Nets to Timed Automata. Electron. Notes Theor. Comput. Sci. 128, 145–160 (2005), doi:10.1016/j.jss.2005.12.021

27. de Castro, L., Timmis, J.: Artificial Immune Systems: A New Computational Approach. Springer, London (2002), `http://www.cs.kent.ac.uk/pubs/2002/1507`
28. Chen, F., Rosu, G.: Towards Monitoring-Oriented Programming: A Paradigm Combining Specification and Implementation. In: Proceedings of the 2003 Workshop on Runtime Verification (RV 2003), Boulder, CO, US (2003), doi:10.1016/S1571-0661(04)81045-4
29. Clarke, E.M., Grumberg, O., Long, D.E.: Model checking and abstraction. ACM Trans. Program. Lang. Syst. 16(5), 1512–1542 (1994), doi:10.1145/186025.186051
30. Clarke, E.M., Grumberg, O., Peled, D.A.: Model Checking. MIT Press (1999)
31. Dasgupta, D., Nino, F.: Immunological Computation: Theory and Applications, 1st edn. Auerbach Publications, Boston (2008)
32. Deb, K.: Multi-Objective Optimization Using Evolutionary Algorithms. Wiley (2001)
33. Dell'Aere, A.: Numerical Methods for the Solution of Bi-level Multi-objective Optimization Problems. HNI-Verlagsschriftenreihe. Heinz Nixdorf Institute, University of Paderborn, Paderborn (2008)
34. Dellnitz, M., Schütze, O., Hestermeyer, T.: Covering Pareto Sets by Multilevel Subdivision Techniques. Journal of Optimization Theory and Application 124(1), 113–136 (2005), doi:10.1007/s10957-004-6468-7
35. Dellnitz, M., Witting, K.: Computation of robust Pareto points. International Journal of Computing Science and Mathematics 2(3), 243–266 (2009), doi:10.1504/IJCSM.2009.027876
36. DeMillo, R.A., Offutt, A.J.: Constraint-based Automatic Test Data Generation. IEEE Transactions on Software Engineering 17(9) (1991), doi:10.1109/32.92910
37. Deuflhard, P., Hohmann, A.: Numerical analysis in modern scientific computing: an introduction, 2nd edn. Springer, New York (2003), doi:10.1007/978-0-387-21584-6
38. Dorociak, R.: Early Probabilistic Reliability Analysis of Mechatronic Systems. In: Proceedings of the Reliability and Maintainability Symposium (2012), doi:10.1109/RAMS.2012.6175464
39. Dorociak, R., Gausemeier, J.: Absicherung der Zuverlässigkeit komplexer mechatronischer Systeme auf Basis der domänenübergreifenden Prinziplösung. In: 25. Fachtagung: Technische Zuverlässigkeit (TTZ), Leonberg, DE (2011)
40. Drusinsky, D.: The Temporal Rover and the ATG Rover. In: Havelund, K., Penix, J., Visser, W. (eds.) SPIN 2000. LNCS, vol. 1885, pp. 323–330. Springer, Heidelberg (2000)
41. Easwaran, A., Kannan, S., Sokolsky, O.: Steering of Discrete Event Systems: Control Theory Approach. Electr. Notes Theor. Comput. Sci. 144(4), 21–39 (2006), doi:10.1016/j.entcs.2005.02.066
42. Eckardt, T., Heinzemann, C., Henkler, S., Hirsch, M., Priesterjahn, C., Schäfer, W.: Modeling and Verifying Dynamic Communication Structures Based on Graph Transformations, pp. 3–22. Springer (2013), doi:10.1007/s00450-011-0184-y
43. Ericson, C.: Hazard Analysis Techniques for System Safety. John Wiley & Sons, Hoboken (2005), doi:10.1002/0471739421
44. Estler, H.C., Wehrheim, H.: Heuristic Search-Based Planning for Graph Transformation Systems. In: Proceedings of the Workshop on Knowledge Engineering for Planning and Scheduling, Freiburg, DE, pp. 54–61 (2011)
45. European Committee for Electrotechnical Standardization (CENELEC): CENELEC EN 50129: 2003. Railway Applications – Communication, Signalling and Processing Systems – Safety Related Electronic Systems for Signalling. European Standard (2003)
46. European Committee for Electrotechnical Standardization (CENELEC): Railway applications Communication, signalling and processing systems Software for railway control and protection systems, CENELEC EN 50128. European Standard (2011)

47. Fenelon, P., McDermid, J.A., Nicolson, M., Pumfrey, D.J.: Towards Integrated Safety Analysis and Design. ACM SIGAPP Applied Computing Review 2(1), 21–32 (1994), doi:10.1145/381766.381770
48. FG Rammig, University of Paderborn: ORCOS – Organic Reconfigurable Operating System, `https://orcos.cs.uni-paderborn.de/doxygen/html` (accessed August 12, 2013)
49. Figueira, J., Greco, S., Ehrgott, M.: Multiple Criteria Decision Analysis: State of the Art Surveys. Kluwer Academic Publishers, Boston (2005), doi:10.1007/b100605
50. Fine, S., Ziv, A.: Coverage Directed Test Generation for Functional Verification Using Bayesian Networks. In: Proceedings of the 40th annual Design Automation Conference, Anaheim, CA, US (2003), doi:10.1145/775832.775907
51. Fischer, T., Niere, J., Torunski, L., Zündorf, A.: Story Diagrams: A New Graph Rewrite Language based on the Unified Modeling Language. In: 6th Int. Workshop on Theory and Application of Graph Transformations (TAGT 1998) (1998)
52. Flaßkamp, K., Heinzemann, C., Krüger, M., Steenken, D., Ober-Blöbaum, S., Schäfer, W., Trächtler, A., Wehrheim, H.: Sichere Konvoibildung mit Hilfe optimaler Bremsprofile. In: Gausemeier, J., Rammig, F.J., Schäfer, W., Trächtler, A. (eds.) Tagungsband zum 9. Paderborner Workshop Entwurf Mechatronischer Systeme, HNI-Verlagsschriftenreihe. Heinz Nixdorf Institute, University of Paderborn, Paderborn (2013)
53. Fox, M., Long, D.: PDDL 2.1: An Extension to PDDL for Expressing Temporal Planning Domains. Journal of Artificial Intelligence Research, 189–208 (2003), doi:10.1613/jair.1129
54. Gausemeier, J., Frank, U., Donoth, J., Kahl, S.: Specification Technique for the Description of Self-Optimizing Mechatronic Systems. Research in Engineering Design 20(4), 201–223 (2009), doi:10.1007/s00163-008-0058-x
55. Gausemeier, J., Rammig, F.J., Schäfer, W. (eds.): Design Methodology for Intelligent Technical Systems. Lecture Notes in Mechanical Engineering. Springer, Heidelberg (2014), doi:10.1007/978-3-642-45435-6_2
56. Geisler, J., Witting, K., Trächtler, A., Dellnitz, M.: Multiobjective Optimization of Control Trajectories for the Guidance of a Rail-bound Vehicle. In: Proceedings of the 17th IFAC World Congress of The International Federation of Automatic Control, Seoul, KR (2008), doi:10.3182/20080706-5-KR-1001.00738
57. Ghallab, M., Nau, D., Traverso, P.: Automated Planning – Theory and Practice. Morgan Kaufmann (2004)
58. Giaquinta, M., Hildebrandt, S.: Calculus of variations. Springer, Berlin (1996)
59. Giese, H., Tichy, M.: Component-Based Hazard Analysis: Optimal Designs, Product Lines, and Online-Reconfiguration. In: Górski, J. (ed.) SAFECOMP 2006. LNCS, vol. 4166, pp. 156–169. Springer, Heidelberg (2006)
60. Gill, P.E., Jay, L.O., Leonard, M.W., Petzold, L.R., Sharma, V.: An SQP Method for the Optimal Control of Large-scale Dynamical Systems. Journal of Computational and Applied Mathematics 120, 197–213 (2000), doi:10.1016/S0377-0427(00)00310-1
61. Gilles, K., Groesbrink, S., Baldin, D., Kerstan, T.: Proteus Hypervisor – Full Virtualization and Paravirtualization for Multi-Core Embedded Systems. In: Schirner, G., Götz, M., Rettberg, A., Zanella, M.C., Rammig, F.J. (eds.) IESS 2013. IFIP AICT, vol. 403, pp. 293–305. Springer, Heidelberg (2013)
62. Groesbrink, S.: A First Step Towards Real-time Virtual Machine Migration in Heterogeneous Multi-Processor Systems. In: Proceedings of the 1st Joint Symposium on System-Integrated Intelligence, Hannover, DE (2012)

63. Groesbrink, S.: Basics of Virtual Machine Migration on Heterogeneous Architectures for Self-Optimizing Mechatronic Systems. Necessary Conditions and Implementation Issues 7, 69–79 (2013)
64. Güdemann, M., Ortmeier, F., Reif, W.: Safety and Dependability Analysis of Self-Adaptive Systems. In: Proccedings of the 2nd International Symposium on Leveraging Applications of Formal Methods, Verification and Validation, ISoLA 2006 (2006), doi:10.1109/ISoLA.2006.38
65. Hagemeyer, J., Hilgenstein, A., Jungewelter, D., Cozzi, D., Felicetti, C., Rueckert, U., Korf, S., Koester, M., Margaglia, F., Porrmann, M., Dittmann, F., Ditze, M., Harris, J., Sterpone, L., Ilstad, J.: A scalable platform for run-time reconfigurable satellite payload processing. In: AHS, pp. 9–16. IEEE (2012), doi:10.1109/AHS.2012.6268642
66. Hagemeyer, J., Kettelhoit, B., Koester, M., Porrmann, M.: Design of Homogeneous Communication Infrastructures for Partially Reconfigurable FPGAs. In: Proceedings of the International Conference on Engineering of Reconfigurable Systems and Algorithms, Las Vegas, NV, US. CSREA Press (2007)
67. Hampton, M., Petithomme, S.: Leveraging a Commercial Mutation Analysis Tool for Research. In: Proceedings of the Testing Academic & Industrial Conference Practice and Research Techniques, Windsor, UK (2007), doi:10.1109/TAIC.PART.2007.39
68. Havelund, K., Rosu, G.: Java PathExplorer – A runtime verification tool. In: Proceedings 6th International Symposium on Artificial Intelligence, Robotics and Automation in Space (ISAIRAS 2001), Montreal, QC, CA (2001), doi:10.1.1.16.1774
69. Henzinger, T.A.: The theory of hybrid automata. In: Proceedings of the 11th Annual IEEE Symposium on Logic in Computer Science, New Brunswick, NJ, US, pp. 278–292. IEEE Computer Society (1996), doi:10.1109/LICS.1996.561342
70. Hillermeier, C.: Nonlinear Multiobjective Optimization – A Generalized Homotopy Approach. Birkhäuser, Berlin (2001)
71. Hölscher, C., Keßler, J.H., Krüger, M., Trächtler, A., Zimmer, D.: Hierarchical Optimization of Coupled Self-Optimizing Systems. In: Proceedings of the 10th IEEE International Conference on Industrial Informatics, Beijing, CN (2012), doi:10.1109/INDIN.2012.6301199
72. Howden, W.E.: Weak Mutation Testing and Completeness of Test Sets. IEEE Transactions on Software Engineering 8(4) (1982), doi:10.1109/TSE.1982.235571
73. International Electrotechnical Commission (IEC): IEC 60812: 2006. Analysis techniques for system reliability – Procedure for failure mode and effects analysis (FMEA). International Standard (2006)
74. International Electrotechnical Commission (IEC): IEC 61025: Fault Tree Analysis (FTA). International Standard (2006)
75. Isermann, R.: Fault-Diagnosis Systems – An Introduction from Fault Detection to Fault Tolerance. Springer, Berlin (2005), doi:10.1007/3-540-30368-5
76. Kerstan, T., Baldin, D., Groesbrink, S.: Full Virtualization of Real-Time Systems by Temporal Partitioning. In: Proceedings of the of the 6th International Workshop on Operating Systems Platforms for Embedded Real-Time Applications, Brussels, BE (2010)
77. Kerstan, T., Oertel, M.: Design of a Real-time Optimized Emulation Method. In: Proceedings of the Design, Automation and Test in Europe, Dresden, DE (2010), doi:10.1109/DATE.2010.5457126
78. King, S., Dunlap, G., Chen, P.: Operating System Support for Virtual Machines. In: Proc. of the USENIX Annual Technical Conference (2003)
79. Kleywegt, A.J., Shapiro, A., Homem-de Mello, T.: The Sample Average Approximation Method for Stochastic Discrete Optimization. SIAM J. on Optimization 12(2), 479–502 (2002), doi:10.1137/S1052623499363220

80. Klöpper, B.: Ein Beitrag zur Verhaltensplanung für interagierende intelligente mechatronische Systeme in nicht-deterministischen Umgebungen. In: HNI-Verlagsschriftenreihe, vol. 253. Heinz Nixdorf Institute, University of Paderborn, Paderborn (2009)
81. Klöpper, B., Aufenanger, M., Adelt, P.: Planning for Mechatronics Systems – Architechture, Methods and Case Study. Engineering Applications of Artificial Intelligence 25(1), 174–188 (2012), doi:10.1016/j.engappai.2011.08.004
82. Klöpper, B., Sondermann-Wölke, C., Romaus, C.: Probabilistic Planning for Predictive Condition Monitoring and Adaptation within the Self-Optimizing Energy Management of an Autonomous Railway Vehicle. Journal for Robotics and Mechatronics 24(1), 5–15 (2012)
83. Koester, M., Luk, W., Hagemeyer, J., Porrmann, M., Rueckert, U.: Design Optimizations for Tiled Partially Reconfigurable Systems. IEEE Transactions on Very Large Scale Integration (VLSI) Systems 19(6), 1048–1061 (2011), doi:10.1109/TVLSI.2010.2044902
84. Kopetz, H.: Real-time systems: design principles for distributed embedded applications. Kluwer international series in engineering and computer science: Real-time systems. Kluwer Academic Publishers (2011), doi:10.1007/978-1-4419-8237-7
85. Kranenburg, T., van Leuken, R.: MB-LITE: A Robust, Light-weight Soft-core Implementation of the MicroBlaze Architecture. In: Proceedings of Design, Automation, and Test in Europe Conference, Dresden, DE (2010), doi:10.1109/DATE.2010.5456903
86. Krüger, M., Witting, K., Dellnitz, M., Trächtler, A.: Robust Pareto Points with Respect to Crosswind of an Active Suspension System. In: Proceedings of the 1st Joint International Symposium on System-Integrated Intelligence, Hannover, DE (2012)
87. Kuhn, H., Tucker, A.: Nonlinear Programming. In: Neumann, J. (ed.) Proceedings of the 2nd Berkeley Symposium on Mathematical Statistics and Probability, Berkeley, CA, US, pp. 481–492 (1951)
88. Lamport, L.: A new solution of Dijkstra's concurrent programming problem. Communications of the ACM 17, 453–455 (1974), doi:10.1145/361082.361093
89. Langseth, H., Portinale, L.: Bayesian Networks in Reliability. Reliability Engineering & System Safety 92(1), 92–108 (2007), doi:10.1016/j.ress.2005.11.037
90. Lee, J., Ni, D., Djurdjanovic, H., Qiu, H., Liao, H.: Intelligent prognostic tools and e-maintenance. Computers in Industry 57, 476–489 (2006), doi:10.1016/j.compind.2006.02.014
91. van Leeuwen, J., Hartmanis, J., Goos, G. (eds.): Partial-Order Methods for the Verification of Concurrent Systems: An Approach to the State-Explosion Problem. Springer, New York (1996), doi:10.1.1.56.8794
92. Leveson, N.G.: Safeware: System Safety and Computers. ACM (1995)
93. Levine, W.: The Control Handbook: Control System Fundamentals, Control System Applications, Control System Advanced Methods. Electrical Engineering Handbook Series. Taylor & Francis Group (2010)
94. Leyendecker, S., Lucas, L.J., Owhadi, H., Ortiz, M.: Optimal control strategies for robust certification. Journal of Computational and Nonlinear Dynamics 5(3), 031,008–031,008 (2010), doi:10.1115/1.4001375
95. Li, J., Zhang, H.C., Lin, Z.: Asymmetric negotiation based collaborative product design for component reuse in disparate products. Computers & Industrial Engineering 57(1), 80–90 (2009), doi:10.1016/j.cie.2008.11.021
96. Luenberger, D.G.: Linear and nonlinear programming, 2nd edn. Addison-Wesley, Reading (1987)

97. Marsden, J.E., West, M.: Discrete Mechanics and Variational Integrators. Acta Numerica 10, 357–514 (2001)
98. Mathew, G., Pinto, A.: Optimal design of a class of hybrid systems with uncertain parameters. In: 50th IEEE Conference on Decision and Control and European Control Conference, Orlando, FL, US, pp. 539–544 (2011), doi:10.1109/CDC.2011.6161357
99. Matzinger, P.: Tolerance, danger, and the extended family. Annual Review of Immunology 12(1), 991–1045 (1994), doi:10.1146/annurev.iy.12.040194.005015
100. Meijer, R.: PDDL Planning Problems and GROOVE Graph Transformations: Combining Two Worlds with a Translator. In: 17th Twente Student Conference on IT (2012)
101. Meyer, T., Keßler, J.H., Sextro, W., Trächtler, A.: Increasing Intelligent Systems' Reliability by using Reconfiguration. In: Proceedings of the Annual Reliability and Maintainability Symposium, RAMS (2013), doi:10.1109/RAMS.2013.6517636
102. Miettinen, K.: Nonlinear Multiobjective Optimization. Kluwer Academic Publishers (1999), doi:10.1007/978-1-4615-5563-6
103. Miner, M.: Cumulative Damage in Fatigue. ASME Journal of Applied Mechanics 12, A159–A164 (1945)
104. Myers, G.J., Sandler, C.: The Art of Software Testing. John Wiley & Sons (2004)
105. National Aerospace Laboratory in the Netherlands: The Safety Methods Database (2012), http://www.nlr.nl/documents/flyers/SATdb.pdf (accessed September 12, 2012)
106. Newcomb, R.W. (ed.): Linear Optimal Control. Networks Series. Prentice-Hall (1971)
107. Ober-Blöbaum, S., Junge, O., Marsden, J.E.: Discrete Mechanics and Optimal Control: An Analysis. Control, Optimisation and Calculus of Variations 17(2), 322–352 (2011), doi:10.1051/cocv/2010012
108. Ober-Blöbaum, S., Seifried, A.: A multiobjective optimization approach for the optimal control of technical systems with uncertainties. In: Proceedings of the European Control Conference, Zürich, CH, pp. 204–209 (2013)
109. Oberthür, S.: Towards an RTOS for Self-Optimizing Mechatronic Systems. In: HNI-Verlagsschriftenreihe. Heinz Nixdorf Institute, University of Paderborn, Paderborn (2010)
110. Object Management Group: Unified Modeling Language (UML) 2.3 Superstructure Specification (2010), http://www.omg.org/spec/UML/2.3/ (Document formal/2010-05-05)
111. Park, R.: Contamination Control, and OEM Perspective. In: Workshop on Total Contamination Control. Centre for Machine Condition Monitoring. Monash University (1997)
112. Peterson, M., Winer, W.: Wear Control Handbook. The American Society of Mechanical Engineers (1980)
113. Pomeranz, I., Reddy, S.M.: On the generation of small dictionaries for fault location. In: Proceedings of the 1992 IEEE/ACM International Conference on Computer-Aided Design, ICCAD 1992, pp. 272–279. IEEE Computer Society Press, Los Alamitos (1992), http://dl.acm.org/citation.cfm?id=304032.304116, doi:10.1109/ICCAD.1992.279361
114. Pook, S., Gausemeier, J., Dorociak, R.: Securing the Reliability of Tomorrow's Systems with Self-Optimization. In: Proceedings of the Reliability and Maintainability Symposium, Reno, NV, US (2012)
115. Popek, G.J., Goldberg, R.P.: Formal Requirements for Virtualizable Third Generation Architectures. Communications of the ACM 17(7), 412–421 (1974), doi:10.1145/361011.361073

116. Porrmann, M.: Adaptive Hardware Platforms for Self-Optimizing Mechatronic Systems. In: International Workshop on Computing in Heterogeneous, Autonomous 'N' Goal-oriented Environments, DAC – Design Automation Conference (2012)
117. Porrmann, M., Hagemeyer, J., Pohl, C., Romoth, J., Strugholtz, M.: RAPTOR–A Scalable Platform for Rapid Prototyping and FPGA-based Cluster Computing, vol. 19. IOS Press (2010), doi:10.3233/978-1-60750-530-3-592
118. Pradhan, D.K. (ed.): Fault-tolerant computer system design. Prentice-Hall, Inc., Upper Saddle River (1996)
119. Priesterjahn, C., Heinzemann, C., Schäfer, W.: From Timed Automata to Timed Failure Propagation Graphs. In: Proceedings of the Fourth IEEE Workshop on Self-Organizing Real-time Systems (2013)
120. Priesterjahn, C., Heinzemann, C., Schäfer, W., Tichy, M.: Runtime Safety Analysis for Safe Reconfiguration. In: IEEE International Conference on Industrial Informatics Proceedings of the 3rd Workshop Self-X and Autonomous Control in Engineering Applications, Beijing, CN, July 25-27 (2012), doi:10.1109/INDIN.2012.6300900
121. Priesterjahn, C., Sondermann-Wölke, C., Tichy, M., Hölscher, C.: Component-based Hazard Analysis for Mechatronic Systems. In: Object/Component/Service-Oriented Real-Time Distributed Computing Workshops, IEEE International Symposium on Object/Component/Service-oriented Real-time Distributed Computing (ISORC), pp. 80–87 (2011), doi:10.1109/ISORCW.2011.19
122. Priesterjahn, C., Steenken, D., Tichy, M.: Component-based timed hazard analysis of self-healing systems. In: Proceedings of the 8th Workshop on Assurances for Self-Adaptive Systems, ASAS 2011, pp. 34–43. ACM, New York (2011), doi:10.1145/2024436.2024444
123. Priesterjahn, C., Steenken, D., Tichy, M.: Timed Hazard Analysis of Self-healing Systems. In: Cámara, J., de Lemos, R., Ghezzi, C., Lopes, A. (eds.) Assurances for Self-Adaptive Systems. LNCS, vol. 7740, pp. 112–151. Springer, Heidelberg (2013)
124. Priesterjahn, C., Tichy, M.: Modeling Safe Reconfiguration with the FUJABA Real-Time Tool Suite. In: Proceedings of the 7th International Fujaba Days (2009)
125. Qanadilo, M., Samara, S., Zhao, Y.: Accelerating Online Model Checking. In: Proceedings of the 6th Latin-American Symposium on Dependable Computing, LADC (2013), doi:10.1109/LADC.2013.20
126. RailCab – Neue Bahntechnik Paderborn: The Project Web Site, `http://railcab.de` (accessed March 5, 2012)
127. Rao, B. (ed.): Handbook of Condition Monitoring. Elsevier (1996)
128. Reinold, P., Nachtigal, V., Trächtler, A.: An Advanced Electric Vehicle for the Development and Test of New Vehicle-Dynamics Control Strategies (2010), doi:10.3182/20100712-3-DE-2013.00172
129. Reutenauer, C.: The mathematics of Petri nets. Prentice-Hall, Inc., Upper Saddle River (1990)
130. Richter, U., Mnif, M., Branke, J., Müller-Schloer, C., Schmeck, H.: Towards a Generic Observer/Controller Architecture for Organic Computing. In: Hochberger, C., Liskowsky, R. (eds.) Tagungsband zur 36. Jahrestagung der Gesellschaft für Informatik – Informatik für Menschen, Dresden, DE. LNI, vol. P-93, pp. 112–119. Bonner Köllen Verlag (2006)
131. Ringkamp, M., Ober-Blöbaum, S., Dellnitz, M., Schütze, O.: Handling High Dimensional Problems with Multi-Objective Continuation Methods via Successive Approximation of the Tangent Space. Engineering Optimization 44(9), 1117–1146 (2012), doi:10.1080/0305215X.2011.634407

132. Röhs, M., Wehrheim, H.: Sichere Konfigurationsplanung selbst-adaptierender Systeme durch Model Checking. In: Gausemeier, J., Rammig, F., Schäfer, W., Trächtler, A. (eds.) Entwurf Mechatronischer Systeme. HNI-Verlagsschriftenreihe, vol. 272, pp. 253–265. Heinz Nixdorf Institute, University of Paderborn, Paderborn (2010)
133. Romaus, C., Bocker, J., Witting, K., Seifried, A., Znamenshchykov, O.: Optimal Energy Management for a Hybrid Energy Storage System Combining Batteries and Double Layer Capacitors. In: Proceedings of the Energy Conversion Congress and Exposition, San Jose, CA, US, pp. 1640–1647 (2009), doi:10.1109/ECCE.2009.5316428
134. Russel, S., Norvig, P.: Artificial Intelligence – A Modern Approach, 2nd edn., pp. 94–136. Prentice Hall (2003)
135. Schütze, O., Witting, K., Ober-Blöbaum, S., Dellnitz, M.: Set Oriented Methods for the Numerical Treatment of Multi-Objective Optimization Problems. In: Tantar, E., Tantar, A.-A., Bouvry, P., Del Moral, P., Legrand, P., Coello Coello, C.A., Schütze, O. (eds.) EVOLVE- A Bridge between Probability. SCI, vol. 447, pp. 185–218. Springer, Heidelberg (2013)
136. Serrestou, Y., Beroulle, V., Robach, C.: Functional Verification of RTL Designs Driven by Mutation Testing Metrics. In: Proceedings of the 10th Euromicro Conference on Digital System Design, Lebeck, DE, pp. 222–227 (2007), doi:10.1109/DSD.2007.4341472
137. Simani, S., Fantuzzi, C., Patton, R.J.: Model-based Fault Diagnosis in Dynamic Systems Using Identification Techniques. Springer, Heidelberg (2002)
138. Slayman, C.: JEDEC Standards on Measurement and Reporting of Alpha Particle and Terrestrial Cosmic Ray Induced Soft Errors. In: Nicolaidis, M. (ed.) Soft Errors in Modern Electronic Systems. Frontiers in Electronic Testing, vol. 41, pp. 55–76. Springer, US (2011), doi:10.1007/978-1-4419-6993-4_3
139. Smith, J.E., Nair, R.: The Architecture of Virtual Machines, vol. 38. IEEE Computer (2005), doi:10.1109/MC.2005.173
140. Sondermann-Wölke, C., Sextro, W.: Integration of Condition Monitoring in Self-Optimizing Function Modules Applied to the Active Railway Guidance Module. International Journal on Advances in Intelligent Systems 3(1&2), 65–74 (2010)
141. Sondermann-Wölke, C., Sextro, W., Reinold, P., Trächtler, A.: Zuverlässigkeitsorientierte Mehrzieloptimierung zur Aktorrekonfiguration eines X-by-wire-Fahrzeugs. In: 25. Tagung Technische Zuverlässigkeit (TTZ 2011) – Entwicklung und Betrieb zuverlässiger Produkte, Leonberg, DE. VDI-Berichte, vol. 2146, pp. 291–302. Düsseldorf (2011)
142. Sterpone, L., Violante, M.: Analysis of the robustness of the TMR architecture in SRAM-based FPGAs. IEEE Transactions on Nuclear Science 52(5), 1545–1549 (2005), doi:10.1109/TNS.2005.856543
143. Tasiran, S., Qadeer, S.: Runtime Refinement Checking of Concurrent Data Structures. In: Proceedings of the 2004 Workshop on Runtime Verification (RV 2004), Barcelona, ES (2004), doi:10.1016/j.entcs.2004.01.028
144. Tichy, M., Klöpper, B.: Planning Self-Adaptation with Graph Transformations. In: Schürr, A., Varró, D., Varró, G. (eds.) AGTIVE 2011. LNCS, vol. 7233, pp. 137–152. Springer, Heidelberg (2012)
145. Timmermann, R., Horenkamp, C., Dellnitz, M., Keßler, J.H., Trächtler, A.: Optimale Umschaltstrategien bei Aktorausfall mit Pfadverfolgungstechniken. In: Gausemeier, J., Rammig, F.J., Schäfer, W., Trächtler, A. (eds.) Tagungsband vom 9. Paderborner Workshop Entwurf mechatronischer Systeme. HNI-Verlagsschriftenreihe. Heinz Nixdorf Institute, University of Paderborn, Paderborn (2013)

146. Trächtler, A., Münch, E., Vöcking, H.: Iterative Learning and Self-Optimization Techniques for the Innovative Railcab-System. In: 32nd Annual Conference of the IEEE Industrial Electronics Society (IECON), Paris, FR, pp. 4683–4688 (2006), doi:10.1109/IECON.2006.347957
147. Tumer, I., Stone, R., Bell, D.: Requirements for a Failure Mode Taxonomy for Use in Conceptual Design. In: Proceedings of the International Conference on Engineering Design, Stockholm, SE (2003)
148. Verein Deutscher Ingenieure (VDI): VDI 2057:2002. Human exposure to mechanical vibrations. Technical Guideline (2002)
149. Vesely, W.E., Goldberg, F.F., Roberts, N.H., Haasl, D.F.: Fault tree handbook – NUREG-0492209. Tech. rep., U.S. Nuclear Regulatory Commission (1981)
150. Wilkinson, P., Kelly, T.: Functional Hazard Analysis for Highly Integrated Aerospace Systems. In: Proceedings of the Ground/Air Systems Seminar (1998), doi:10.1.1.28.8417
151. Witting, K.: Numerical Algorithms for the Treatment of Parametric Multiobjective Optimization Problems and Applications. In: HNI-Verlagsschriftenreihe. Heinz Nixdorf Institute, University of Paderborn, Paderborn (2011)
152. Witting, K., Ober-Blöbaum, S., Dellnitz, M.: A Variational Approach to Define Robustness for Parametric Multiobjective Optimization Problems. Journal of Global Optimization (2012), doi:10.1007/s10898-012-9972-6
153. XILINX: MicroBlaze Processor Reference Guide, V9.0 (2008)
154. Zhao, Y., Rammig, F.: Online Model Checking for Dependable Real-Time Systems. In: 15th IEEE International Symposium on Object/Component/Service-Oriented Real-Time Distributed Computing (ISORC), pp. 154–161. IEEE Computer Society, Shenzhen (2012), doi:10.1109/ISORC.2012.28
155. Zilberstein, S.: Using Anytime Algorithms in Intelligent Systems. AI Magazine 17(3), 73–83 (1996), doi:10.1.1.41.3559

Chapter 4
Case Study

Abstract. The challenge of increasing the dependability of a self-optimizing system is best addressed by a development process which seamlessly integrates appropriate dependability-oriented methods. The developer is assisted during this process by a set of guidelines to help him or her select the most suitable methods for the current development step.

In this chapter, a development process making use of these guidlines for the effective development of a self-optimizing system is shown exemplarily and advantages are highlighted that can be obtained by expanding the development process to include methods of increasing the dependability of the resulting system . Furthermore, the interaction between and sequence of these methods is outlined. As a case study, the RailCab system is presented and an abstract of its development process is illustrated.

In general, the development process is structured into the two phases: "Conceptual Design" and "Design and Development". During Conceptual Design of the Active Guidance Module of the RailCab, methods based on the specification of the Principle Solution are employed to being increasing the dependability as early as possible. This is followed by the Design and Development phase, during which methods of analyzing the dependability of the whole RailCab system, as well as of optimizing the system behavior, are employed.

In Sect. 3.3, a methodology was presented, which supports the improvement of the dependability of self-optimizing systems. Its constituent elements are a method database, a guide for planning the use of the selected methods in the development process, and the appropriate software tool. By using the presented methodology, the developer can decide easier and faster which of the vast number of available dependability engineering methods (like shown for self-optimizing systems in Chap. 3) suits the development task best. In the following, the use of this methodology in an actual development project (the RailCab mentioned in previous chapters) will be discussed as a concrete application of the strategies presented thus far in the book. Additionally, the methods, selected by engineering teams to improve the dependability of the RailCab will be discussed in detail.

J. Gausemeier et al. (eds.), *Dependability of Self-optimizing Mechatronic Systems*, Lecture Notes in Mechanical Engineering,
DOI: 10.1007/978-3-642-53742-4_4, © Springer-Verlag Berlin Heidelberg 2014

4.1 Selecting Suitable Methods Using the Methodology

Rafal Dorociak, Jürgen Gausemeier, Peter Iwanek, Tobias Meyer, and Walter Sextro

As mentioned above, the development of self-optimizing systems is divided into two main phases: the (domain-spanning) conceptual design and the (domain-specific) design and development. Within the conceptual design phase, the principle solution is developed by using the specification technique CONSENS (Sect. 2.1). After the specification of the domain-spanning models, these models are used as the first inputs to analyze the system's dependability (Chap. 2.1). There are certain methods that can be used for this analysis from which the developer may choose (e.g. the *Early Probabilistic Reliability Analysis* (cf. Sect. 3.1.1)). In the domain-specific design and development phase, more models are created by the experts from the different domains (e.g. models specified in MECHATRONICUML or MCAD). Based on these domain-specific models, further analyses can be performed. In the following, we will show the support available to a developer when using the methodology to search for and select suitable dependability methods for particular tasks (cf. Sect. 3.3), using the railway system RailCab as an example.

In our example, we show how an engineer can search for suitable methods of improving the dependability of a system. Figure 4.1 displays a diagram excerpt of the development process.

In this figure, the bubbles show where the methodology is used to choose suitable dependability engineering methods according to the specific development task. At this point, the safety engineer has just started to search explicitly for dependability methods, so he has selected the following search criteria: "conceptual design" for the development phase, "rail engineering" for the industry sector and the CENELEC 50128 [4] standard, which is the industrial standard for the rail engineering industry. Based on the chosen criteria, the irrelevant methods are filtered out and displayed in the graphical user interface (GUI) of the software tool[13] (Fig. 4.2).

Three methods are proposed, all of which conform to the recommendations of the CENELEC 50128 standard and can be conducted during the conceptual design: 1) early FMEA (Failure Mode and Effects Analysis) based on the principle solution, 2) early FTA (Fault Tree Analysis) based on the principle solution and a combination of both early integrated FMEA and FTA based on the principle solution. The engineer can now select the *early FMEA based on the principle solution* (1) for use, which is based on the principle solution of the subsystem Active Guidance Module. With this method the critical weak points of the system can be identified and respective counter-measures can be defined; the principle solution will be more reliable. The safety engineer is provided with additional information regarding the proposed method in form of a method description, which is shown in the GUI. Figure 4.3 shows a section of the method description for the early FMEA method. The method description contains relevant information, for example the summary, the required

[13] The graphical user interface shown in Fig. 4.2 is a prototypical version of the software tool. What is important for the usage of the methodology is the core concept behind the software (e.g. classification of dependability engineering methods, etc.).

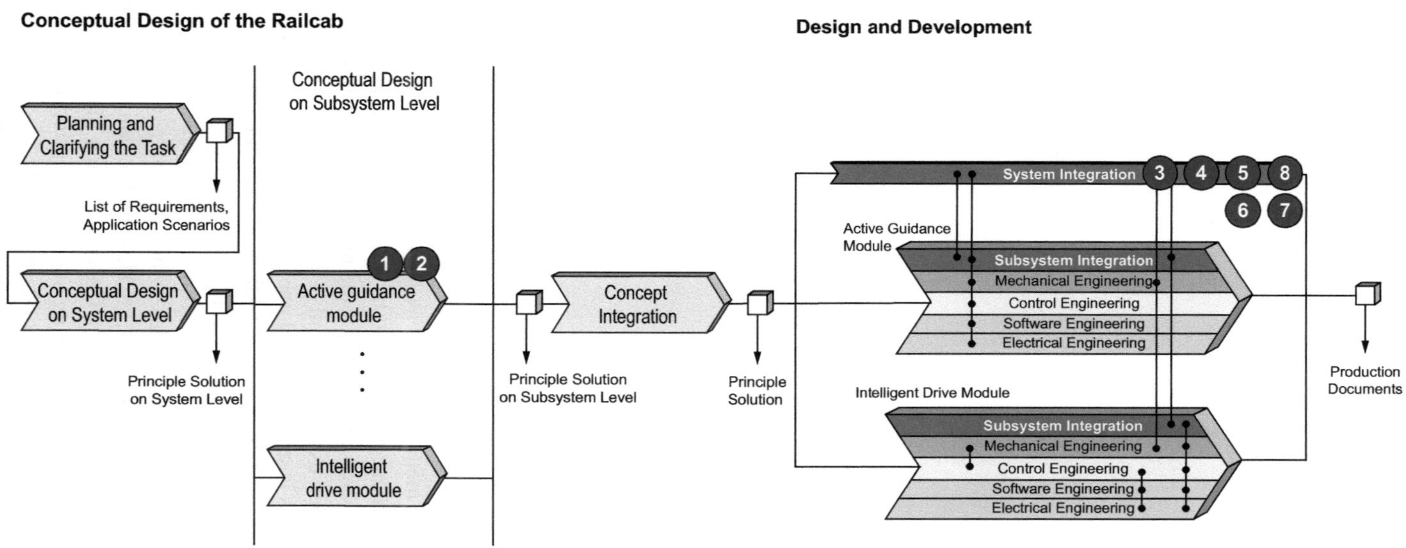

Bubble	Name	Section
1	Early Probabilistic Reliability Analysis of an Advanced Mechatronic System based on its Principle Solution	3.1.1
2	Early Design of the Multi-Level Dependability Concept	3.1.2
3	Verification with Motion Profiles for Interacting Mechatronic Systems	3.2.10
4	Self-Healing in Operating Systems	3.2.12
5	Safe Planning	3.2.9
6	Behavior Planning	3.2.5
7	Behavioral Adaption of Differing Model Parameters	3.2.7
8	Virtualization	3.2.15

Fig. 4.1 Development Process and related methods

Fig. 4.2 The search function of the method database

input could be the result of the method, as well as some supporting documents. These documents could be for instance, phases and milestones diagrams, templates for the usage of the method, and exemple usages of the method (e.g. older projects where the method has already been used successfully). In addition, the information concerning phases and milestones is stored in the database as a formal model which can be used for the generation of a dependability workflow, as explained in Sect. 3.3.

After using early FMEA the engineer found that the reliability of the system could end up too low and thus not fulfill the given requirements. The engineer runs another search using the methodology and decides to use the *early design of the Multi-Level Dependability Concept* (2) (Figure 4.1). This method allows an early implementation of the Multi-Level Dependability Concept, which allows a self-optimizing system to react to its own deteriorating state by adapting the system behavior, thus increasing one or several attributes of dependability, i.e. the system's reliability. After the phase "Concept Integration", the development process is split

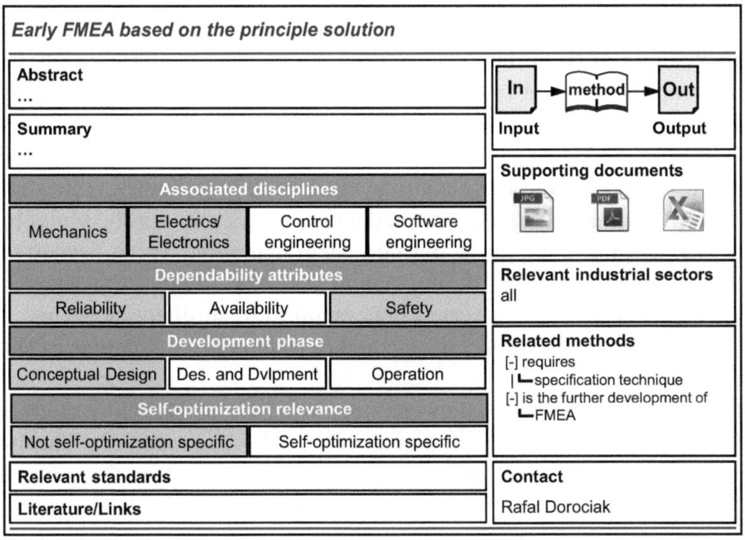

Fig. 4.3 The method description for the early FMEA based on the principle solution (excerpt)

up into several parallel development domains (mechanics, electrics/electronics etc.) and the system integration. In the system integration phase, the safety engineer recognizes that varying braking capabilities of several RailCabs in a convoy could lead to severe accidents. Thus, a solution must be found and implemented. To do so, the safety engineer can use the *verification with motion profiles for interacting mechatronic systems* (3), which is excellently suited to this task. This method identifies the braking capabilities and defines safety conditions for them, allowing the engineer to eliminate the possibility of accidents from this error source.

In order to obtain dependable software, a model-based development approach using MECHATRONICUML was selected, within which the software architecture and communication behavior of several RailCab vehicles could be modeled. It was then found that failures in certain components could lead to severe faults, necessitating a resistant system. To design this system, the method *Analysis of Self-healing behavior* (4) is selected and used. To allow for online adaptation of the system's communication structure, the software architecture of the RailCab system has to able to be reconfigured dynamically. Such reconfigurations can be modeled using the story pattern formalism of MECHATRONICUML. In order to decide *when* and *how* to reconfigure the architecture of the system at runtime, an integrated planning system is used, which selects appropriate reconfigurations and decides upon their ordering. The resulting sequence of reconfigurations is not permitted to produce unsafe configurations, which is why the method *Safe Planning* (5) must be employed. This method takes safety requirements under consideration while searching for a given target configuration.

At the same time, other engineering groups which are evaluating the dynamic behavior of the vehicle system can employ the method *Behavior Planning* (6). During operation, this method proactively generates plans for selecting the overall sequence of optimal solutions from the Pareto sets of multiobjective optimization. In addition, a probabilistic model is used to determine deviations from the deterministic plan. These, deviations between the actual and the planned system behavior, unfortunately, cannot always be avoided and may have a multitude of causes. Deviations may be caused by unquantifiable environmental influences; however, they may also be caused by deviations between the model parameters used for the multiobjective optimization and those measured during system operation. These deviations have to be detecteted and corrected during system operation in order to achieve system dependability. For this purpose, the safety engineer, control engineers and mathematicians in the team can employ the method *Behavior-Based Adaption of Differing Model Parameters* (7).

The last method, which the safety engineer will use together with experts in software engineering during the development phase, is *Virtualization* (8). The dynamic mapping of virtual machines onto physical platforms increases the runtime flexibility of the system software architecture, which can be exploited to improve the reliability and availability of software components. By applying virtual machine migration, a software component can potentially be kept functional despite a possible hardware failure. In case of a partial memory failure or the breakdown of a coprocessor, the input/output relationship with the component can be shifted to another unit. The execution of the VM (virtual machines) is suspended and the executable code, memory content, and processor state are communicated to the target processor, where the execution is resumed. Virtual machine migration is therefore a technique for controlling hazards. In the next section, the usage of the methods mentioned here will be described in more detail.

4.2 Development of the Active Guidance Module

Rafal Dorociak, Jürgen Gausemeier, Peter Iwanek, Tobias Meyer, Walter Sextro, and Christoph Sondermann-Woelke

The development process, see also Fig. 4.1, is subdivided into two major phases. During the first phase, the whole system is modularized and then a conceptual design for the individual modules is developed. The RailCab has been subdivided into several modules all of which serve separate primary functions. Among these is the Active Guidance Module, which has already been introduced in Sect. 1.3.1.2. The use of methods chosen using the guidelines as shown in Sect. 4.1 will be examined here in detail.

4.2.1 Early Probabilistic Reliability Analysis of an Advanced Mechatronic Systems Based on Its Principle Solution

First, the early FMEA is performed based on the specification of the Principle Solution. In order to do so, the entire RailCab system has to be specified, using the

Failure Mode and Effects Analysis (FMEA)
module: servo-cylinder

system element	function	failure mode	failure effect	s	failure cause	d	o	rpn	counter or detection measure	...
hydraulic valve regulation	regulate position of the valve	hydraulic valve regulation provides no switch position for the 4/4-way valve	valve does not change the pressure on the output anymore	6	servo cylinder regulation does not provide to-be valve slider position	2	7	84	monitor the outgoing communication towards the 4/4-way valve	...
					hydraulic valve regulation is broken	9	3	162	generate a warning message, if needed	
					energy supply of the hydraulic valve regulation is interrupted	7	6	252	monitor energy supply	
4/4-way valve	close the valve	valve closes in an unwanted manner	hydraulic valve stays in the current position	8	energy supply is interrupted	7	4	224	monitor energy supply	...
					magnetic coil is damaged	8	3	192	generate a warning message, if needed	
servo-cylinder regulation	determine the as-is value for the cylinder lifting way	a wrong as-is value for the cylinder lifting way has been determined	control deviation $e = X^*_{cylinder} - X_{cylinder}$ has a wrong value; the desired lifting way will not be accomplished	9	sensor is damaged	8	3	216	monitor as-is value of the cylinder lifting way $X_{cylinder}$	...
					cable break	9	2	162		
					sensor has been wired in a wrong way	3	2	54	redundant design of the measuring system	
					wrong calibration	2	5	90		
...	...	...	...	...						

s: severity of the failure effect
d: detection probability of the failure cause
o: occurence probability of the failure cause
rpn: risk priority number (rpn = s*d*o)

Fig. 4.4 FMEA of the Active Guidance Module (excerpt)

specification technique CONSENS (cf. Sect. 2.1). For each system element, all possible failures, their possible causes and the consequences are described in tabular form. For the identified causes and consequences, the corresponding detection and counter-measures are determined and also recorded in the FMEA table. The FMEA table has to be filled out manually, which usually happens in moderated workshops with a safety engineer and experts from the involved domains. The counter-measures are usually proposed and developed by the respective developers. A subtask of FMEA is risk prioritization by assigning a the risk priority number (RPN, cf. [1,8]). By using the RPN, dependability-critical system elements can be determined. Figure 4.4 shows a section of the FMEA table of the Active Guidance Module. For example, we see that the actuator can be damaged, the flange could break or the RailCab could derail.

According to the workflow generated by the software tool (cf. Sect. 4.1), the phases of the *early FTA based on the Principle Solution* are executed next [2]. The specification of the Principle Solution (mainly the partial model Active Structure) is expanded to include the description of the failure propagation; the underlying procedure is based on the FTA method. Figure 4.5 shows a section of the specification of the failure propagation within the Active Guidance Module. It shows, in particular, that the failure of the eddy-current sensor causes a failure of the actuator, because the actuator cannot receive any control values; the consequence is a complete failure

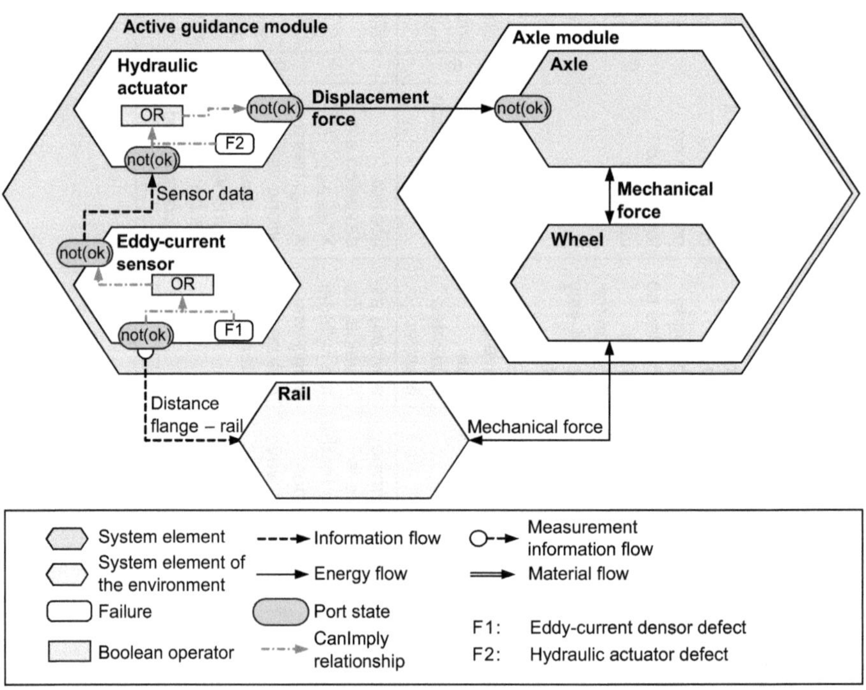

Fig. 4.5 FTA of the Active Guidance Module (excerpt)

of the track guidance module. A complete failure of the track guidance is critical for the dependability, particularly when driving onto a passive switch. In this case, the risk of derailing is very high. To prevent this scenario, mechanical track guidance can be installed and a railway extension is attached next to the passive switches. The mechanical track guidance switches to the desired side before passing a passive switch. If the active steering breaks down, the RailCab is led by an interlocked connection in the given direction [3, 5].

Using the early integrated FMEA and FTA, weak points in the system with regard to dependability can be found and respective counter-measures can be implemented before critical failure states are reached. Both methods are carried out based on the domain-spanning description of the Principle Solution. Thus, especially fault propagation paths which cross domain boundaries can be analyzed. Further information about the usage of the early FMEA and the FTA based on the Principle Solution in combination with the methodology described here (c.f. Sect. 3.3) can be read in [3].

4.2.2 Early Design of the Multi-Level Dependability Concept

To increase the dependability of the Active Guidance Module during runtime, the Multi-Level Dependability Concept, see Sect. 3.2.1, has been chosen. This method has been developed to enable a self-optimizing system to actively adapt its behavior according to the current system deterioration. To do so, pre-calculated optimal workings points that offer better dependability than those in use are selected during runtime. This method is used to include the Multi-Level Dependability Concept at an early design phase.

To utilize the method *Early Design of the Multi-Level Dependability Concept*, input from several sources is required. As has been explained in Sect. 3.1.2, the main source of information is the Principle Solution. The Principle Solution is comprised of several partial models that each contain information about one area of interest. Using the development process outlined in Sect. 1.2.3, it is the main result of the Conceptual Design Phase. Additionally, information obtained by completing an FMEA is required. It is therefore only natural at this point to employ the method *Early Probabilistic Analysis of an Advanced Mechatronic System based on its Principle Solution* (see Sect. 3.1.1). This method itself is based on information included in the Principle Solution as well and thus requires no further sources of information.

For the calulation of the required optimal working-points, multiobjective optimization is used. Thus, this method heavily relies on the availability of multiobjective optimization and its required foundations. It is also necessary to ensure that additional dependability-related objectives can be included in the optimization model. The method *Reliability-Oriented Multiobjective Optimization*, see Sect. 3.2.11, assists the developer by analyzing the effects of objectives intended to increase the dependability.

All of the results from this method are embedded into the Principle Solution, which serves as input for the phase "Design and Development". By carrying this part of development out in the "traditional" way, i.e. separated into domains, all components of the Multi-Level Dependability Concept are implemented just like other system components. These in turn require other methods to ensure dependable system behavior, which are not specific to the implementation of the Multi-Level Dependability Concept.

4.3 Development of the RailCab Vehicle

Kathrin Flaßkamp, Stefan Groesbrink, Philip Hartmann, Christian Heinzemann, Bernd Kleinjohann, Lisa Kleinjohann, Martin Krüger, Sina Ober-Blöbaum, Claudia Priesterjahn, Christoph Rasche, Wilhelm Schäfer, Dominik Steenken, Ansgar Trächtler, Heike Wehrheim, and Steffen Ziegert

Following the conceptual design of the individual modules, the resulting concepts must be integrated and the second phase of the development process, the Design and Development, is begun. This phase consists of individual design and development processes for each module. They are conducted individually in each domain using the results from the Conceptual Design Phase, the Principle Solution, as specifications. If domain-spanning partial models have been engineered, the results must be synchronized.

Parallel to the development of the individual modules the results of the domains are integrated into the overall system model in the system integration. These span all of the modules and take the full system into account. During this phase and in these processes, methods are one more chosen using the guidelines as described in Sect. 4.1.

4.3.1 Verification with Motion Profiles for Interacting Mechatronic Systems

In Sect. 3.2.10, we gave an overview of a method combining several tools, processes and methods to create a set of software solutions that would bridge the gap between continuous and discrete behavior. As an example, we used the proposed RailCab system. In this section, we will examine how that method works in the broader context of the design process of the RailCab system and how it interacts with other design and dependability methods.

Prior to employing this method, the required inputs first need to be generated. For the RailCab system, this means the physical parameters of the RailCab need to be determined first. This applies to the outer shape of the RailCab (or at least an abstracted box to represent the boundaries) and any physical parameters that have an influence on the performance of actuators, such as vehicle mass.

Furthermore we need a clear idea of the behavior the RailCab will exhibit in the finished system. That means we need to know all the different situations a RailCab may encounter, e.g. single drive, convoy drive, track switching, entering and exiting stations and so on. Each of these situations will need a motion profile, as will any transition between them.

The final piece of information for the continuous part of the method are the goal functions, with regard to which the motion profiles must be optimized. These are part of the overall optimization concept and should have been fixed earlier in the development process *D.M.f.I.T.S*, [6], Sect. 3.6. Based on these pieces of information, motion profiles will be generated for the RailCabs, relying on the techniques of optimal control, *D.M.f.I.T.S*, [6], Sect. 5.3.6.

In constructing the communication protocol that will assign these motion profiles at runtime, we must rely on a pre-existing communication topology. This is provided by the modeling of the system in MECHATRONICUML *D.M.f.I.T.S*, [6], Sect. 5.2. Here we find communication topologies for all the different situations specified above, as well as basic protocol implementations that do not yet take the motion profiles into account.

Based on the existing communication topology, we expand the existing communication protocols by adding profile distribution protocols and compatibility checks as described in Sect. 3.2.10. These protocols are then verified using design-time verification *D.M.f.I.T.S*, [6], Sect. 5.2.3.

As a result, we get MECHATRONICUML models of the commmunication protocols guiding the RailCab that have been made aware of the physical dimensions and characteristics of the vehicles and the resulting safety restrictions. These protocols can then be used in the remaining process just as the previous MECHATRONICUML models would have been, including their eventual implementation in actual code. Once in place, the modifications of the communication protocols produced using this method interlock with the artefacts created by several other methods.

For safe planning (see Sect. 4.3.3), the new protocols introduce an added layer of complexity, since actions that are planned to take place in the future may be rejected by the inherent safety check now implemented in the communication protocols. For example, a convoy merge that was planned may be rejected by the target convoy because there is no combination of motion profile choices that satisfies the safety requirements.

The control engineers creating the actuator controllers that will actually drive the RailCab must be made aware of the motion profiles that have been constructed, as they will need to support these motion profiles in the Controller.

4.3.2 Analysis of Self-healing Operations

Verification identifies systematic faults, i.e., faults in the system design. It does not consider random faults that occur, e.g. due to the wear of hardware parts. Random faults may lead to hazards, which in turn may lead to accidents, such as the collision of two RailCabs in a convoy. These faults cannot be eliminated completely. However for a dependable system like the RailCab, the developer has to guarantee that hazards can only occur within an acceptable probability.

Should a hazard occurrence probability be too high, the system design must be changed in order to reduce this probability. One means for the reduction of hazard occurrence probabilities is self-healing. The self-healing system detects an error at runtime and reacts by initiating a self-healing operation that removes the error from the system or stops it from propagating to a critical point in the system. However, the hazard occurrence probability is only reduced if the self-healing operation is executed quickly enough. To guarantee this, the analysis of self-healing operations is employed as presented in Sect. 3.2.8.

The analysis of the self-healing operations requires information from earlier design phases as input: the thresholds for hazard occurrence probabilities of the system are provided by the Requirements, which result from the Domain-spanning Conceptual design (cf. Sect. 2.1). MECHATRONICUML as described in *D.M.f.I.T.S*, [6], Chap. 5, provides models which specify the system architecture, as well as the state-based and reconfiguration behavior.

The analysis of self-healing operations uses the behavior models to generate failure propagation models with timing information which are needed to analyze the propagation times of failures. The system architecture and the reconfiguration rules which specify self-healing operations are needed to analyze how the self-healing operation affects the propagation of failures in the system architecture.

Before the self-healing operations are analyzed, the system models need, of course, to be verified. We apply the verification approach which is explained in *D.M.f.I.T.S*, [6], Chap. 5.

The hazard analysis according to Giese et al. [7] is a prerequisite and part of our analysis of self-healing operations. During the analysis of self-healing operations, it is used to compute hazard occurrence probabilities before and after the self-healing; thus, the developer can assess whether the self-healing operation has reduced the hazard occurrence probability according to the safety requirements.

During safe planning, the results of the online hazard analysis must be taken into account. If a transition is locked that was part of a safe plan, the safe plan needs to be adapted accordingly.

4.3.3 Safe Planning

To reconfigure the communication structure of the RailCab system, the method *Safe Planning* is used, cf. Sect. 3.2.9. This method allows runtime reconfigurations to be planned by an integrated software system. It also obeys safety requirements, i.e. it prevents the system from reconfiguring into dangerous configurations.

MECHATRONICUML introduces *real-time coordination patterns* that represent communication protocols between subsystems, cf. *D.M.f.I.T.S*, [6], Sect. 5.2.1. A real-time coordination pattern specifies the behavior of subsystems as real-time statecharts and determines their communication link. These real-time coordination patterns can be activated and deactivated dynamically by runtime reconfigurations.

The application scenario in Sect. 3.2.9.4 uses the activation and deactivation of specific real-time coordination patterns during a RailCab's movement as an example, e.g. the story pattern *cConvoy* to operate in convoy mode or *cPublication* to register at a base station. A runtime reconfiguration specifying the creation of a *Convoy* real-time coordination pattern affects not only the system's software; to operate in convoy mode, a RailCab changes its subcomponents' configurations and thus gives a different subcomponent control over the velocity of the RailCab. While planning these runtime reconfigurations can be done on a discrete level, abstracting aspects of the subcomponents away, the actual execution of these runtime reconfigurations necessarily involves these subcomponents. One of these subcomponents is the RailCab's hybrid energy storage system, cf. *D.M.f.I.T.S*, [6], Sect. 2.1.5. This system also applies a planning technique for its internal behavior and an online adaptation of this behavior, e.g. to decide how much energy to invest in the passengers' comfort, cf. the methods *Behavior Planning* and *Behavior-Based Adaptation of Differing Model Parameters* in Sect. 3.2.5 and 3.2.7. Its output constitutes a single Pareto point for each track segment. The system continuously measures sensor data and ensures that its intended operation mode will always be met by recalculating the Pareto point during operation if necessary.

Another subcomponent integrates the hazard analysis of self-healing behavior presented in Sect. 3.2.8. It specifies runtime reconfigurations to be executed when a specific error is detected. In principle, it is possible to associate a target configuration instead of a runtime reconfiguration with an error or failure, and plan a sequence of runtime reconfigurations that reduces the hazard probability with the method *Safe Planning*. A similar approach is used in [11]. Applied to the RailCab system, this means that the method *Safe Planning* can be used at two different levels of abstraction: to plan the de-/instantiation of real-time coordination patterns during a RailCab's movement in the Cognitive Operator of the OCM and to compute self-healing plans in the Reflective Operator of the OCM, cf. Sect. 1.1.1. The latter was not done for the RailCab system because the employed planning system would have had to provide runtime guarantees in order not to invalidate the timing predictions made by the hazard analysis.

4.3.4 Behavior Planning

By using the Method *Behavior Planning* (cf. Sect. 3.2.5), the RailCab has to choose from different behaviors (operation modes) on each section of track. This choice represents a trade-off between several objective functions (e.g. minimization of the body movement or energy losses). Because of the complexity of solving multiobjective optimization problems, this process is done during the design phase of the RailCab system for each track section. During operation, the RailCab chooses solutions from the Pareto sets.

In the case of the RailCab, the state space is a combination of the operation modes and the state of charge (SOC) of the on-board energy storage. Planning procedures such as state space search algorithms explore this state space in order to define the operation mode to be executed for each track section. Of course, the availability of the operation modes depends on the current state of charge, and thus the planning procedure should choose them in such a way that a given objective function (e.g. minimizing the total energy consumption) is optimized across all track sections [9].

During operation, deviations from the deterministic plan are caused by unquantifiable not-exactly-quanitifiable (sounds odd, though) influences (e.g. wind speed or rain fall). To build up a probabilistic model of environmental influences and the energy transfer resulting from the operation modes, Bayesian networks have to be defined for the RailCab-domain at design time. For each operation mode, a corresponding Bayesian networks describes the conditional probability distribution $P(\Delta_{SOC}^{i}|\varepsilon_i)$ with Δ_{SOC}^{i} denoting the change in the state of charge and ε_i the relevant environmental influences. An inference made through the Bayes Networks can then determine the probability distribution in the change of state of charge Δ_{SOC}^{i} from the probability distribution of the environmental influences (cf. [10]).

4.3.5 Behavior-Based Adaptation of Differing Model Parameters

The method *Behavior-Based Adaptation of Differing Model Parameters* as introduced in Sect. 3.2.7 is used to increase the dependability of self-optimizing systems during movement. The method has been developed to enable systems which create plans based on Pareto points to adapt themselves to environmental changes.

The method should be utilized if several unexpected conditions can occur during operation to which the system has to react. The system under consideration has a restricted energy supply and has to travel with the available energy to its arrival station. During travel, the passengers expect a certain degree of comfort, which has a direct influence on the expected energy consumption, resulting in conflicting requirements. As these two requirements are only a part of all requirements, a multiobjective optimization process, as presented in Sect. 3.2.6, has to be performed.

Based on the resulting Pareto points, computed by the method *Computation of Robust Pareto Points* as described in Sect. 3.2.6, it is possible to create a plan (cf. *D.M.f.I.T.S*, [6], Sect. 5.3.8), consisting of several Pareto points, which leads the RailCab from its initial position to its destination. This plan has to provide a sequence of safe configurations, as described in Sect. 3.2.9.

Nevertheless, a plan computed offline does not take into account the actual operating conditions of the RailCab. One example might be strong headwind, leading to an increase in power consumption. Such an event can make it impossible for the RailCab to reach its arrival station before it runs out of energy.

Another use case for which this method should be selected is if convoy travel is possible but has not been considered by the planner during offline planning. In this case, the energy consumption decreases for all RailCabs except the first one. This makes it possible to expend more energy on, for example, passenger comfort. In both cases, the model Pareto fronts might not take the current situation into account. It is then necessary to compute new points in order to find a feasible plan.

A method such as the *Behavior-Based Adaptation of Differing Model Parameters* can overcome the described problem. It takes the initially computed plan and compares the computed Pareto point currently being used to determine the RailCab properties with a Pareto point computed from currently measured data. To take measuring errors into account, a threshold is implemented. If the difference between the

Pareto point selected using the initial plan and the Pareto point computed using the measured data is above this threshold, an adaptation of the Pareto front used for the initial plan is conducted.

The measured points must not necessarily be Pareto points; hence, it might not be possible to compute a new Pareto front which includes the measured point. If that is the case, the Pareto front given by the model parameters will be approximated such that it approaches the measured point.

The resulting Pareto points are delivered to the planner, which takes them into account for replanning. This leads directly to a self-optimizing behavior based on the current system state. The method reacts quickly to changing influences and allows the behavior-based adaptation of the system before it is too late to react, such as reacting to an unexpected energy demand before it becomes impossible to reach the arrival station.

4.3.6 Virtualization

Virtualization as introduced in Sect. 3.2.15 is utilized in order to increase both run-time flexibility and dependability of the system software architecture. Virtual machine migration improves two major attributes of dependability, namely reliability and availability. It enables a coarse-grained dynamic load balancing and can increase the robustness by fostering fault resilience in case of hardware failures, such as partial memory failures, a breakdown of I/O if there is an I/O connection to other processing elements remaining, or the failures of a coprocessor. If the hardware failure does not preclude the saving and transfer of the virtual machine state, the operation of the subsystems on the affected electronic control unit can be continued on another unit. Virtual machine migration is an additional technique for controlling hazards, and consequently reduce the likelihood that the hazards pose a threat to life, health, property or environment. Failures such as the breakdown of the power supply or the central processing unit usually prevent a transfer of code and data to another processing element. In some cases, the system can benefit from self-diagnosing hardware that signals upcoming hardware failures on the basis of built-in self-tests, and can then use this information to perform the migration before it becomes impossible.

In the Operator Controller Module architecture for self-optimizing systems (see Sect. 1.1.1), virtualization can be applied to the components of the Cognitive Operator and the components of the Reflective Operator. The OCM includes timing requirements for the applications on all of its three levels. The actuator-controlling software on the lowest level and the supervising Reflective Operator on the level above have to meet hard real-time requirements. Virtualization is an integration approach of coarse granularity and introduces both an execution time and a memory overhead, which is inappropriately high for low-level controllers with significantly higher real-time requirements. This holds true especially for migration and the involved downtime of the migrated component. It is feasible for components on the higher levels, but much less suited for actuator-controlling low-level software with its high execution frequencies.

The self-optimization procedures of the Cognitive Operator have to fulfill soft real-time requirements and a violation of the timing requirements does not result in a safety-critical malfunctioning of the mechatronic system. The self-optimization applications of the Cognitive Operator execute the most computationally intensive algorithms. There are often execution mode alternatives with differing resource demands, associated with different optimization levels. Switching between them at runtime leads to varying resource consumption, which the flexible resource assignment of virtualization and migration can be applied to adapt. A migration manager has to decide *which* virtual machine should be migrated *where* and *when*. Advanced planning approaches should be applied in order to take into account smart migration decisions based on the current resource utilization and the timing constraints of the virtual machines, such as the *Safe Planning* introduced in Sect. 3.2.9.

References

1. Birolini, A.: Reliability Engineering – Theory and Practice, 5th edn. Springer, Heidelberg (2007), doi:10.1007/978-3-662-03792-8
2. Dorociak, R.: Early Probabilistic Reliability Analysis of Mechatronic Systems. In: Proceedings of the Reliability and Maintainability Symposium (2012), doi:10.1109/RAMS.2012.6175464
3. Dorociak, R., Gaukstern, T., Gausemeier, J., Iwanek, P., Vaßholz, M.: A Methodology for the Improvement of Dependability of Self-Optimizing Systems. Production Engineering – Research and Development 7(1), 53–67 (2013), doi:10.1007/s11740-012-0425-3
4. European Committee for Electrotechnical Standardization (CENELEC): Railway applications Communication, signalling and processing systems Software for railway control and protection systems, CENELEC EN 50128. European Standard (2011)
5. Gausemeier, J., Rammig, F.J., Schäfer, W. (eds.): Selbstoptimierende Systeme des Maschinenbaus. HNI-Verlagsschriftenreihe, vol. 234. Heinz Nixdorf Institute, University of Paderborn, Paderborn, DE (2009)
6. Gausemeier, J., Rammig, F.J., Schäfer, W. (eds.): Design Methodology for Intelligent Technical Systems. Lecture Notes in Mechanical Engineering. Springer, Heidelberg (2014), doi:10.1007/978-3-642-45435-6_2
7. Giese, H., Tichy, M.: Component-Based Hazard Analysis: Optimal Designs, Product Lines, and Online-Reconfiguration. In: Górski, J. (ed.) SAFECOMP 2006. LNCS, vol. 4166, pp. 156–169. Springer, Heidelberg (2006)
8. International Electrotechnical Commission (IEC): IEC 61025: Fault Tree Analysis (FTA). International Standard (2006)
9. Klöpper, B., Aufenanger, M., Adelt, P.: Planning for Mechatronics Systems – Architecture, Methods and Case Study. Engineering Applications of Artificial Intelligence 25(1), 174–188 (2012), doi:10.1016/j.engappai.2011.08.004
10. Klöpper, B., Sondermann-Wölke, C., Romaus, C.: Probabilistic Planning for Predictive Condition Monitoring and Adaptation within the Self-Optimizing Energy Management of an Autonomous Railway Vehicle. Journal for Robotics and Mechatronics 24(1), 5–15 (2012)
11. Tichy, M., Klöpper, B.: Planning Self-Adaptation with Graph Transformations. In: Schürr, A., Varró, D., Varró, G. (eds.) AGTIVE 2011. LNCS, vol. 7233, pp. 137–152. Springer, Heidelberg (2012)

Chapter 5
Conclusion and Outlook

Tobias Meyer, Claudia Priesterjahn, and Walter Sextro

Self-optimizing mechatronic systems offer capabilities well beyond those of traditional technical systems. They are able to react autonomously to changing system objectives based on an evaluation of their current situation. Their development requires the joint effort of several domains: mechanical engineering, electrical engineering, control engineering, software engineering and mathematics, and their advanced functionality makes self-optimizing mechatronic systems complex a challenge to develop. This challenge is taken up by a custom-tailored development process which is introduced in the accompanying book "Design Methodology for Intelligent Technical Systems" [1].

However, self-optimizing mechatronic systems require high dependability, as they are often used in safety-critical environments or are subject to high demands on availability. In this book, the development process of [1] has been expanded to include methods that enable the developer to guarantee or increase the dependability of a self-optimizing mechatronic system. Guidelines for the selection of these methods is proposed and all necessary selection criteria are given; this should enable the developer to quickly select a suitable method for his or her current engineering problem.

In Chap. 1, self-optimizing mechatronic systems were introduced to the reader. Self-optimization not only offers expanded capabilities, but also poses challenges in particular the challenge of how to ensure the dependability of such systems. Here, we have presented our definition of dependability and its application to self-optimizing systems. A brief introduction to our proposed development process, which is the subject of the accompanying book "Design Methodology for Intelligent Technical Systems" [1], is given. One main challenge arises due to the multidisciplinary approach that is required to develop a self-optimizing system, for which we have also suggested certain coordination strategies to overcome the difficulties inherent here. Throughout this book, several demonstrators were used to exemplify different aspects of self-optimizing systems introduced in the chapters. Among these is the RailCab, a system of innovative rail vehicles that offer individual transport for freight or passengers.

J. Gausemeier et al. (eds.), *Dependability of Self-optimizing Mechatronic Systems*,
Lecture Notes in Mechanical Engineering,
DOI: 10.1007/978-3-642-53742-4_5,

In Chap. 2, we presented the parts of the development process which focus on dependability. Based on the development process, we have presented a methodology which helps the developer of a self-optimizing system ensure its dependability. The information in this chapter should help in deciding whether the system's dependability needs to be increased further and offers methods of doing so.

Chapter 3 presents methods which focus on guaranteeing and improving the dependability of self-optimizing mechatronic systems. These methods cover multiple domains which are involved in the development of these systems. All of the method descriptions follow the same structure in order to facilitate their application. Their goal, the prerequisites, a detailed description of their application and the results are given sequentially for each section. Additionally, the methods contain real-world examples using the demonstrators which were introduced in Chap. 1. Each method also contains the necessary criteria.

In Chap. 4, we apply the methodology and the methods presented in Chaps. 2 and 3 to the RailCab. We have improved parts of the RailCab that did not fulfill dependability requirements using the methods described within this book. The RailCab is now a dependable self-optimizing mechatronic system.

The methods which have been presented in this book were developed in the course of the Collaborative Research Center 614. Methods that do not focus on self-optimizing systems, such as standard methods, are thus outside the scope of this book. They have to be incorporated into the development process presented here while it is being used in the development of new products.

The RailCab system and all other demonstrators used in this book are academic research projects. To fully evaluate the suitability of these methods and the accompanying guidelines, an evaluation based on industrial development processes and the problems arising in that environment might prove worthwhile.

Reference

1. Gausemeier, J., Rammig, F.J., Schäfer, W. (eds.): Design Methodology for Intelligent Technical Systems. Lecture Notes in Mechanical Engineering. Springer, Heidelberg (2014), doi:10.1007/978-3-642-45435-6_2

Index

MIX
Papier aus verantwortungsvollen Quellen
Paper from responsible sources
FSC® C105338

If you have any concerns about our products,
you can contact us on
ProductSafety@springernature.com

In case Publisher is established outside the EU,
the EU authorized representative is:
Springer Nature Customer Service Center GmbH
Europaplatz 3, 69115 Heidelberg, Germany

Printed by Libri Plureos GmbH
in Hamburg, Germany